HEAVY EQUIPMENT

JOHN TIPLER

CHARTWELL
BOOKS, INC.

Published by
CHARTWELL BOOKS, INC.
A Division of **BOOK SALES, INC.**
114 Northfield Avenue
Edison, New Jersey 08837

Copyright © 2000 Amber Books Ltd

All rights reserved. No part of this publication may be reproduced, stored in a retrieval system or transmitted, in any form or by any means, electronic, mechanical, photocopying, recording or otherwise, without the prior written permission of the copyright holder.

ISBN: 0-7858-1171-0

Amber Books Ltd
Bradley's Close
74-77 White Lion Street
London N1 9PF

Editor: Helen Wilson
Design and computer artwork: Richard Burgess
Picture research: Lisa Wren

Printed in Italy

ACKNOWLEDGEMENTS

I should like to thank the following people for their help while writing and compiling this book. My old friend Gerald Stubbs – Stubbsy – recounted numerous anecdotes from his time at the controls of many of the machines described here; I have included Stubbsy's accounts of how to drive a bulldozer, excavator, dumper and motor grader, and they enhance the text immeasurably. I also want to credit Dr Paul Nieuwenhuis, heavy equipment expert from Cardiff Business School, internet maestro Andy Robinson, JCB owner Mark Tomczynski, Canadian photographer Mike Schram and Zoë and Alfie Tipler for Docklands inspiration. Manufacturers who kindly supplied material for inclusion included Adrian Wilkinson, Lincolnshire County Archives, Mike Fraser & Franz Josef Bentgens of Krupp Materials Handling Ltd, Carol Nicholson of Bucyrus Europe, Sayuri Oka of Kawasaki, Japan, Chris Davis of Interface Marketing, Peter Winkel and Merilee Hunt of Liebherr, Bill Williams of P&H Mining, Andrew M. Grobengieser of New Holland North America, Inc., Simon Bridge of Orenstein & Koppel Ltd (O&K) and Bryan Hawickhorst of Caterpillar Media Management Systems.

CONTENTS

Introduction	6
Excavators	24
Mining, Machinery, Tunnel Borers and Drilling Rigs	50
Bulldozers	64
Motor Scrapers	76
Motor Graders	92
Wheeled Loaders and Dozers	102
Dump Trucks	118
Cranes	142
Road Rollers, Compactors and Tarmac Layers	160
Index	174

INTRODUCTION

Since time immemorial, we've been modifying our environment for one reason or another. In ancient times we built monuments like the Pyramids of Egypt and Central America and defensive bastions like the Great Wall of China, largely without the help of mechanical technology. More recently, we've been able to construct protective barriers like the levees along the banks of major waterways such as the Mississippi, and created the polders of Holland out of land that would otherwise be below sea level.

To contain water for our cities, we build vast reservoirs behind dams like the Hoover, Aswan and the current Three Gorges project in China. We have created canal systems such as those that weave their way across Europe, and move mountains in our quest for fossil fuels and precious metals. In the late twentieth century we needed ever-larger airports, such as the complicated construction at Hong Kong, and other communications systems like the railway networks and motorways. All these have called upon mechanical equipment in their construction.

The business of constructing our cities alone required quarrying technology to source the raw materials for building, and this gave us the lime, sand and stone quarries that grew so large that they were worked out and new ones came on stream. All these activities have constituted

The NASA Space Shuttle is moved to its launch site on dual crawler transporters known as Hans and Franz, each weighing almost 2700 tons and forming one of the largest pieces of heavy equipment in the world. The rocket-powered shuttle may be supersonic, but the crawler platform goes at just 1.6km/h (1 mile/h).

INTRODUCTION

One of the world's most impressive construction sites is the Chek Lap Kok airport, completed in 1998 and built entirely on a 1248-hectare (3083-acre) man-made island to the west of Hong Kong, which it serves.

major earth-moving tasks. But huge international works like the Kiel, Suez and Panama canals were relatively recent compared with ancient Chinese achievements – their Grand Canal built around 800AD was 966km (600 miles) long, and by 984AD they had locks on their waterways.

Advancing mechanisation in the nineteenth century made this sort of activity possible on an increasingly expansive scale, while the equipment needed to perform the tasks grew commensurately larger and more specialised. Other areas demanding special equipment include mining, tunnelling, bridge and road building, farming and forestry, not to mention the servicing of military ambitions. Yet heavy equipment technology has advanced beyond all recognition since the first steam-powered excavators ground into action in the mid-nineteenth century and superseded the wheelbarrows, picks and shovels of the navvies responsible for many of the backbreaking tasks in canal and railway building.

There are several main categories of heavy plant and equipment, which are themselves composed of a number of variations on a theme, and some of which have overlapping functions. At one end of the scale we have mining machines and tunnel borers burrowing deep below the surface. In open-cast mines and quarries we find walking-dragline excavators and bucket-wheel excavators, with stripping shovels, bulldozers, scrapers and graders in the middle, all of which are allied to massive dump trucks for transporting the product. An array of road-building equipment is on hand to make the highway surface, while an equally diverse range of cranes is available for the construction of houses and tower blocks. At a more domestic level, we use tractor-based products typified by the JCB.

The ubiquitous JCB comes in all sizes and has many applications, from building site to farmyard, from waterways clearance to materials handling depot. My relatives in Cumbria, northern England, have a JCB 3C that they used for underpinning the old house they were restoring. It transpired that it was built on clay, with little in the way of foundations, and the JCB's backhoe proved ideal for scooping out below the walls so that concrete could be poured in. Being self-sufficient types, they even used the JCB to dredge gravel from their local river to produce the aggregate for the concrete. One trick that Tom, its operator, evolved, was to pivot the whole machine over a stream by jacking it up on its front shovel and rear backhoe, with the front bucket dug in on one side and the rear bucket on the

INTRODUCTION

other bank. He was then able to arch the JCB body from one side of the river to the other by shoving it forward with the backhoe. To get an idea of the versatility of the JCB, you need to witness the demonstration by the company's display team, known as the Dancing Diggers. This consists of ten full-on JCBs that perform all kinds of synchronised stunts including a Mexican wave, a roll-over, and a backhoe tunnel that allows a vehicle to pass under them. Impressive stuff, and you can catch them at county shows, not necessarily just in the UK.

BIG IS BEAUTIFUL

In the closing years of the twentieth century, items of heavy equipment were among the most jaw-dropping and awesome mobile machines in use on the planet, such was their size and stature. From giant dumpers, each as large as a dozen trucks, to gargantuan bucket-wheel excavators working the open-cast coalmines, these leviathans dwarf the everyday vehicles that we routinely use in urban areas, yet these are the machines that created and, at a distance, sustain our domestic environment. This book sets out to illustrate just how far we've come, showing what the state-of-the-art manufacturers are producing now, and tracing how they achieved their present position.

Dump trucks are a fundamental tool of the construction industry, and come in all shapes and sizes. This is a Caterpillar 769, introduced in 1962, rugged and dependable and viewed with something close to affection by the workers who drove it.

CAT 769

Make: *Caterpillar*
Model: *769 rear-dump hauler*
Manufactured: *1962*
Engine: *Cat D348 V12 turbodiesel*
Power output: *960bhp*
Carrying capacity: *35 tons*

STEAM POWER

In the early days of mechanisation, motive power was still provided by mules and oxen, which hauled the early scoops and blades until self-propelled diggers became the norm. The first exploratory steps into the potential of steam power were taken in the early 1830s by the American William Smith Otis, who patented his steam-powered bucket shovel in 1836. A pair of Otis steam shovels were operational in Brooklyn's Prospect Park in 1842, while another was used on the construction of the Western Railroad in 1858 at Springfield, Massachusetts. Its 1.14-cubic metre (1.5-cubic yard) capacity was said to be the equal of no fewer than 80 men. During the decades that followed, steam-powered machines played a major role in the industrialised nations of Europe and North America in urban development, as well as in mining operations and the building of canals and railways. By 1884 the first fully revolving multi-bucket power shovels were made in the UK by Whitaker & Sons, rotating on a continuous-belt principal. Some of these machines were constructed to run on rails, with more track being laid down as the project progressed, and these rails could also provide the means of disposing of the waste matter. Among the first applications was in the mining industry, where mechanised digging

Electric stripping shovels are colossal diggers, and this Bucyrus 495-BI is the company's largest. Launched in 1996, it can claw 42.8 cubic metres (56 cubic yards) of overburden weighing up to 85 tons from a mine or quarry face with each bucketful.

increased productivity and relieved manpower of certain arduous tasks.

The machines were also particularly useful in canal excavation. The Suez Canal that ran between the Mediterranean and the Red Sea was built under the direction of Ferdinand de Lesseps' International Suez Canal Company between 1859 and 1869. It was 161km (100 miles) long and deep enough for ocean-going ships, and the British Lincoln-based Ruston & Dunbar company's steam excavators proved invaluable. Ruston was the first company to capitalise on the steam excavator, certainly outside the USA, in association with partners that included Dunbar, Proctor and Burton.

Equally well documented was one of the last canals to be built during the British Industrial Revolution, the Manchester Ship Canal. At 58km (36 miles) long, it was perhaps less impressive than the Egyptian waterway in terms of distance, but nevertheless it connected Manchester with the sea and provided a significant trading route during the first half of the twentieth century. It was opened in 1894, having taken up to 17,000 labourers and some 97 steam-powered excavators, eight dredgers, 124 cranes, 192 portable engines and 212 pumps. The excavators, including one manufactured by Ruston & Proctor of Lincoln, ran on temporary railway lines. The Ruston & Proctor machine weighed 35.84 tons and was capable of excavating 1529 cubic metres (2000 cubic yards) in the course of a 10-hour day, while smaller versions were made by J.H. Wilson & Company of Liverpool and Whitakers of Horsforth, Yorkshire, with a capacity

INTRODUCTION

of perhaps 300 to 400 cubic metres (392 to 524 cubic yards) a day. The men who built the canals were known as Navvies – short for Navigators – and the excavators were soon correspondingly known as Steam Navvies. In their wake came the offerings from Ohio-based Bucyrus and the Marion Steam Shovel Company, which both became major manufacturers of excavators. Other major projects included the daunting Panama Canal between the Atlantic and Pacific oceans, which although only 64km (40 miles) long, took two goes to finish. De Lesseps' first attempt was stymied in 1889 by the Malaria and Yellow Fever that killed 22,000 of his workforce, and the canal was finally completed by the US Army in 1919 after being dogged by revolution and engineering problems. Around 45,000 labourers were employed on the Panama Canal, as well as numerous steam shovels which shifted an estimated 184 million cubic metres (240 million cubic yards) of material.

WALKING DRAGLINES

Next innovation in the excavation arena was the dragline, which is an excavator operated by a complex system of cables to dig and hoist earth from beneath the level it stands on. The concept was patented in Chicago by John W. Page in 1904, and as it stands the largest piece of heavy equipment in the world today is the 168-cubic metre (20-cubic yard)

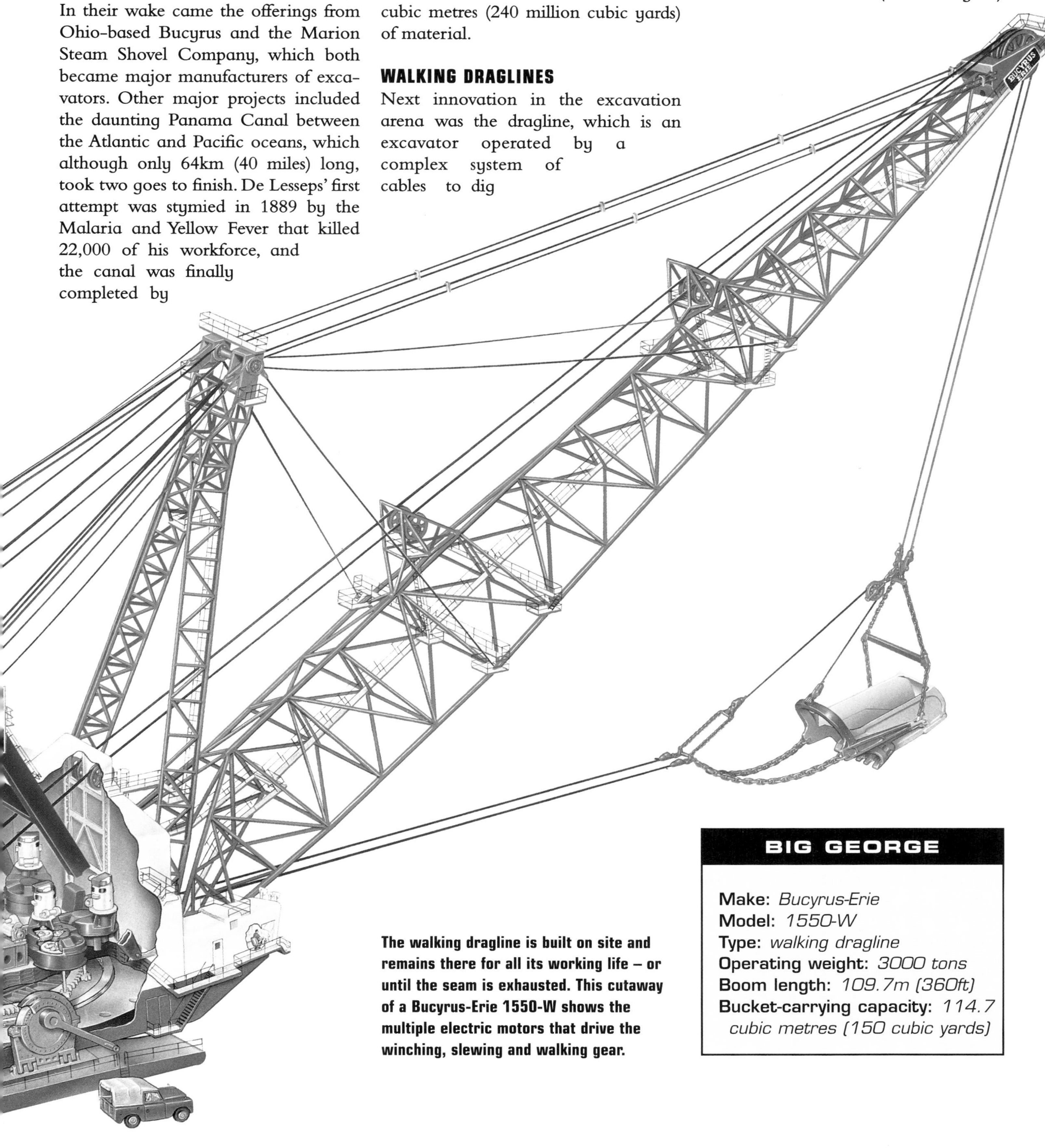

The walking dragline is built on site and remains there for all its working life – or until the seam is exhausted. This cutaway of a Bucyrus-Erie 1550-W shows the multiple electric motors that drive the winching, slewing and walking gear.

BIG GEORGE

Make: *Bucyrus-Erie*
Model: *1550-W*
Type: *walking dragline*
Operating weight: *3000 tons*
Boom length: *109.7m (360ft)*
Bucket-carrying capacity: *114.7 cubic metres (150 cubic yards)*

Trenching machines like this Vermeer Manufacturing Topcon behave like giant chainsaws, carving trenches out of the ground with their boom-mounted continuous blade, tipped with a multitude of carbide cutters that rip into the earth and rock. Booms can be up to 9m (30ft) long.

capacity bucket of 'Big Muskie', a 12,244-ton dragline excavator built by Bucyrus-Erie for American Electric Power in 1969 and operational in open-cast coal mining in Ohio. From around 1911, power shovels were equipped with tracks – bands of loosely linked steel slats that ran in a continuous band over a set of wheels driven by steam-powered pistons or, later, by petrol-fuelled engines. At a stroke this provided greater mobility as well as stability, and about this time the forerunners of today's capacious bucket-chain and bucket-wheel excavators appeared in Northern Europe, while the first large stripping shovels and walking draglines were operational in quarrying, excavating and open-cast mining. Walking draglines are related to crawler cranes, except that, to put it simply, one is an excavator and the other is a hoister.

Pioneering powered excavator and dragline manufacturers that set up in North America in the last two decades of the nineteenth century included the Automatic Shovel Company, Osgood & MacNaughton, Bucyrus Foundry & Manufacturing Company, and the Marion Steam Shovel Company. The introduction of electric motors and internal combustion engines created new drivetrains for excavators, and the first to use a petrol engine was the Monighan dragline of 1910. By the 1920s it was becoming customary to mount excavators on crawler chassis, and as early as 1921 Bucyrus had developed a diesel-powered shovel. Most truck makers didn't entertain diesel power until the 1930s, and it wasn't universally used until the 1950s. In the 1920s, excavator manufacturers of varying sizes included Priestman and Smith, Ransomes & Rapier, and Ruston & Hornsby in the UK, with Link-Belt, Northwest, P&H, Manitowoc, and Lima in the USA. In Germany there was Demag, Åkerman in Sweden, and in Japan, Kobe and Hitachi were operational. Ruston & Hornsby merged with Bucyrus-Erie in 1930 to form Ruston Bucyrus, and the initial product of the alliance was the 10-RB of 1934, which remained in production until 1969, with nearly 8000 units built. Equally long-running was the 22-RB introduced in 1937 and still in production in the late 1980s.

MINING SHOVELS

The most heavy-duty excavators in the post-war period were cable-operated mining shovels. During the 1940s and 1950s a large number of manufacturers sprang up making cable-operated excavators in the main, but the advent of efficient hydraulic systems rendered them largely redundant. Draglines, however, are of necessity cable-operated machines, and are still very much in use today. Manufacturers include P&H, FMC-Link-Belt, Koehring, Manitowoc, and Northwest.

In the world of hydraulic-operated excavators, among the leaders were the French company Poclain, which produced its first hydraulic excavator in 1951, and came out with its popular TY45 model in 1956. Another French company, SICAM, patented a hydraulic excavator in 1954 called the Yumbo 525. Over in neighbouring Germany, hydraulic excavators were manufactured by Atlas, Demag, and Liebherr. As with trucks – and indeed elsewhere in the motor industry – the 1960s was a time of expansion, with many smaller firms merging and products being manufactured under licence by others. In 1961, Mitsubishi began producing the Yumbo Y35 and H25 under licence. These were the first hydraulic excavators in Japan, although it would take another decade before Mitsubishi was building its own machines, beaten by Komatsu and Hitachi.

INTRODUCTION

In 1966, the somewhat aptly named Menzie Muck was the first climbing hydraulic backhoe. Made in Switzerland, it was sufficiently versatile for operations such as ski-lift construction. At the small-scale end of the market, vehicles like JCB's 802 mini-excavator of 1983 had a bucket capacity of 0.1 cubic metre (an eighth of a cubic yard), while at the other end of the scale, O&K's RH300 of 1979 had a capacity ranging from 22.2 to 33.6 cubic metres (29 to 44 cubic yards). Then again, only three examples of this colossal 498-ton hydraulic mining shovel were produced – each powered by two diesel engines or electric motors delivering 2352bhp – while smaller machines like the JCB item proliferate. By far the most common excavator is the simple universal shovel unit, invariably mounted on a caterpillar-tracked chassis, powered by a petrol or diesel engine, and capable of being fitted with a variety of attachments for tackling specific tasks.

Advancing civilisation needs feeding of course, and to an extent the agricultural revolution became mechanised at the same rate as the urban industrial revolution. Central to the rapid development of the twentieth-century farm and construction industries was the humble tractor, which took the form of a traction engine prior to the invention and adoption of the internal combustion engine. The innovators of the late nineteenth century applied tracks to steam traction engines to create vehicles that could operate in soft sand or certainly in off-road sites that would have more conventional vehicles foundering. Among the first companies to experiment with tracked propulsion was Hornsby, who had the world's first track-laying tractor up and running in 1896. Probably the first commercially successful crawler tractor was introduced in 1904 by Benjamin Holt who, with Daniel Best, went on to found what would become the giant Caterpillar concern. Holt's crawler tractors were thereafter widely used in construction sites such as the Los Angeles Aqueduct project of 1908, and during that muddy bloodbath known as the First World War literally thousands of Holt crawlers were pressed into service. Fitted with a front-mounted blade, the crawler was christened the bulldozer.

In a civilised country you don't get far without a viable road system, and

In the late 1990s, O&K's RH400 was the largest hydraulic excavator in the world, and machines like this were challenging the big cable-operated stripping shovels and draglines that were traditionally noted for an enormous appetite for soil.

there was clearly a niche for more specialised earth-moving machines. Thus in 1912, the Galion Company was producing its tractor-hauled grader, and by 1915 the Russell Grader Company had come out with the first self-propelled grader. This piece of kit was to play a major role in the road construction industry and is still fundamentally the same machine used today to level undulations and shape and angle the earth in preparation for the new road's foundations and metalled surfaces. The first fully integrated large-capacity scraper was introduced in 1922 by R.G. LeTourneau. A scraper performs a more fundamental job than a grader, in so far as it excavates as it goes along instead of simply redepositing the earth.

Wheeled loaders like this Bell L2208-C at ISCOR's Newcastle-upon-Tyne slag-handling site, play a crucial role in mining operations, transferring material from quarry face to dump truck for deposition.

Scrapers can be self-propelled or tractor-powered, and by 1940 LeTourneau self-propelled scrapers were in active use building airstrips to aid the Allied war effort around the world.

WHEELED LOADERS

An important category of earth-mover emerged as an offshoot of the regular tractor in the 1920s, and this was the wheeled loader. Originally, tractors were simply fitted with buckets – as they still are at a farmyard level – and broader buckets came to be fitted as hydraulics came into wider use, along with the backhoe version that appeared in the late 1930s. This type of vehicle is generally seen as a JCB product, or at its most heavy duty, made by Caterpillar. The most commonly used types have a bucket capacity of around 0.76m (one cubic yard) and are powered by 80bhp diesel engines, while at the top end of the range come the 6.1-cubic metre (8-cubic yard) bucket loaders, which commonly use 700bhp engines. In exceptional cases, some front-end loaders exceed 22.9-cubic metre (30-cubic yard) capacity. Some Caterpillar wheeled loader applications are timber, rock and waste handling, feats which use various types of bucket or grabbing apparatus. It's a simple step from the wheeled loader to the tracked loader. This derivative is mounted on crawler tracks in order to be operational in situations inaccessible to the wheeled loader.

As with tracked excavators, wheeled loaders come in an array of differing sizes, from massive quarrying vehicles such as those made by LeTourneau and International Hough to small domestic units. They also have a wide range of fittings, from regular buckets of varying shapes and capacities to cutters and clippers, claws, grapples, jackhammers, tree-stump removers and trenching tools.

The first fully-integrated wheeled

INTRODUCTION

Almost all earth-moving projects commence with the removal of topsoil, and the most efficient machine for this operation is a scraper like this Terex unit, which can form the earth into heaps.

loader, complete with rubber tyres, was the Hough HS Payloader of 1939. The company, founded by Frank G. Hough, later introduced the hydraulically-controlled bucket for wheeled loaders, as well as four-wheel drive, torque converters and hydrostatic transmission. Unsurprisingly, given the acquisitive nature of companies in the automotive industry and associated fields, Hough was bought by International Harvester and went on to become part of Komatsu Dresser. Next phase in the history of wheeled loaders came in 1945, when Caterpillar offered its Traxcavator front-end loader shovel. This device could be fitted to its tractors, and the bucket was operated by cable and pulley, and simply lowered by gravity. J.C. Bamford started out in 1945, and the JCB vehicle that became used to define the wheeled loader concept, like Hoover did for vacuum cleaner, was introduced in 1954. The wheeled loader and backhoe combination was launched in 1957 by J.I. Case in its Model 320, and a further evolution on the theme was the three-wheel loader developed by the E.G. Melroe Company and marketed as the Bobcat skid-steer loader from 1958. Such was its popularity that by 1990 the US market accounted for some 27,000 skid-steer loaders. Virtually all the manufacturers involved in heavy plant construction have a finger in the wheeled loader market at one level or another. You'll find everyone there, from Caterpillar, Daewoo TCM, Mitsubishi, Melroe and LeTourneau, to Liebherr, Kubota, Komatsu-Dresser, Kobelco, JCB, John Deere and J.I. Case.

DUMP TRUCKS

There's no point having an excavator if you've got nothing to move the earth around in, and the excavator's natural ally is the dump truck. Although dump trucks predate loaders by several

Thundering along the haul road at 56km/h (35 miles/h) is this Komatsu HD785 100-ton rear-dump hauler. The life of the dumper driver can be tedious, going repeatedly from quarry face to tip site and back again.

decades, the tandem pair form a familiar partnership at construction sites. Dump trucks range in size more greatly than loaders, from 200-ton giants serving mining excavators to the more common 50-ton construction truck, with a range of off-highway sizes between. Early trucks had to be adaptable and versatile and were pressed into a variety of duties. Among the foremost producers of the more specialised dump truck was Mack, which came out with the AC Bulldog model in 1916 and the AP in 1926. Most truck manufacturers have a construction site

INTRODUCTION

Bulldozers can be fitted with different blade configurations according to the type of material they are pushing. This Cummins-powered Liebherr PR752 Litronic tractor is heaping up light topsoil.

line in their vehicle range, but a few makers have specialised in a particular field – Dennis and MAN have been noted for their fire engines, while Oshkosh's speciality is snow-clearance vehicles and concrete carriers. But in the world of heavy-duty earth movers, it's firms like Caterpillar, Euclid and Terex who are the stars. These trucks can easily be as big as a two-storey house, such is their capacity for carting and delivering material at quarry sites, and the driver needs his own staircase to access the cab.

Probably the first off-road dump truck was Euclid's TracTruk of 1934, and as mining and excavation equipment grew ever larger, dump trucks expanded in stature to keep pace. The advent of hydraulics made it possible for the idiosyncratic dumper body to carry and tip large volumes of matter. Back in 1958, Western's 150-ton Hauler was the benchmark dump truck, while the world's largest dumper, the 356-ton Terex Titan, made its appearance in 1971. This gargantuan truck was powered by a 3300bhp engine mated to an alternator that drove four electric motors mounted in the rear axles. It served in an open-cast coalmine at Sparwood, British Columbia, and it remains unique, as only the one was built. More recently, the accent on dump-truck design has been toward greater articulation and manoeuvrability. For example, Donson Equipment's articulated DDT-630 of 1995 deposits its cargo by ejecting it with a hydraulic ram. Other modern features include electronic monitoring systems, faster operating hydraulics, and self-cleaning brakes.

BULLDOZERS AND SCRAPERS

Elsewhere in the chain of earth-moving vehicles we find bulldozers, scrapers and graders. The expression to bulldoze was coined in the 1870s and meant to intimidate someone to get them to submit. But in mechanical terms, the bulldozer concept was introduced in the First World War, when the off-road potential of tracked vehicles like the first tanks became apparent. Front-mounted blades became regular fitments in the 1920s, for snow clearance as well as earth moving. By the Second World War, such vehicles were in regular use for site clearance, hauling scrapers and graders, and as power and size have increased, their modes of operation have kept pace with the ongoing introduction of hydraulics, electronics and safety features. The world's biggest blade belonged to the Dubble Dude Blazer, measuring 14.6m (48ft) wide and operated by a pair of crawler tractors. It was built in 1971 by Bladerson Inc., and could shift 10,842 cubic metres (14,180 cubic yards) an hour. Major manufacturers today include John Deere, Caterpillar, and Komatsu Dresser, whose D575 A-2 ranks as the largest single-blade bulldozer in the world, capable of moving 68.8 cubic metres (90 cubic yards) in a single push. Caterpillar's D11N model is the largest US-made bulldozer, weighing in at 97,524kg (215,000lb) and powered by a 770bhp V8 motor. An amphibious version made by Komatsu is used to dredge sand and shingle from the sea-bed to improve harbour access

Caterpillar is the most prolific producer of crawler bulldozers, and its D10 model caused a sensation when released in 1977 because of its elevated sprocket drive, installed to reduce wear on moving parts.

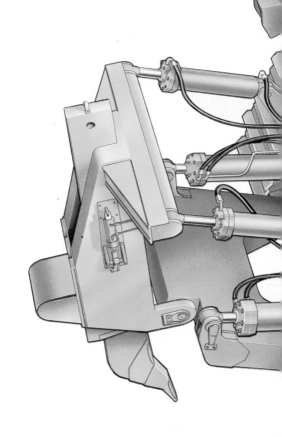

INTRODUCTION

and clear channels for shipping.

While the bulldozer's function, right from its origins as a crawler tractor with a mounted blade, has been to push and grade earth, the scraper's job is to shave off high ground and fill in low-lying depressions. The grader follows on by smoothing out and angling the ground, usually in preparation for road-building. Graders and scrapers can be self-propelled or tractor-pulled, and some scrapers are amongst the largest earth movers, with capacities for shifting 38.2 cubic metres (50 cubic yards) of matter.

Indeed, the honours for the largest scraper go to LeTourneau's LT-360, which was powered by eight engines producing 5080bhp. Known as the Electric Digger, it consisted of three linked blades, with total hopper capacity of 360 tons.

By no means a new concept, back in the early nineteenth century horse-drawn wagons with under-slung blades were used as graders. One of these devices took the concept a stage further. It was the 'Little Wonder' of 1885, invented by J.D. Adams, in which the wheels were made to lean in order to provide stability and counter the sideways thrust of its angled blade. The Little Wonder concept was developed further by the Adams Company and the Russell Grader Company in the

CAT D10 BULLDOZER

Make: *Caterpillar*
Model: *D10*
Manufactured: *1977*
Engine: *Cat 3412 turbodiesel*
Power output: *520bhp*
Blade width: *5.2m (17ft)*

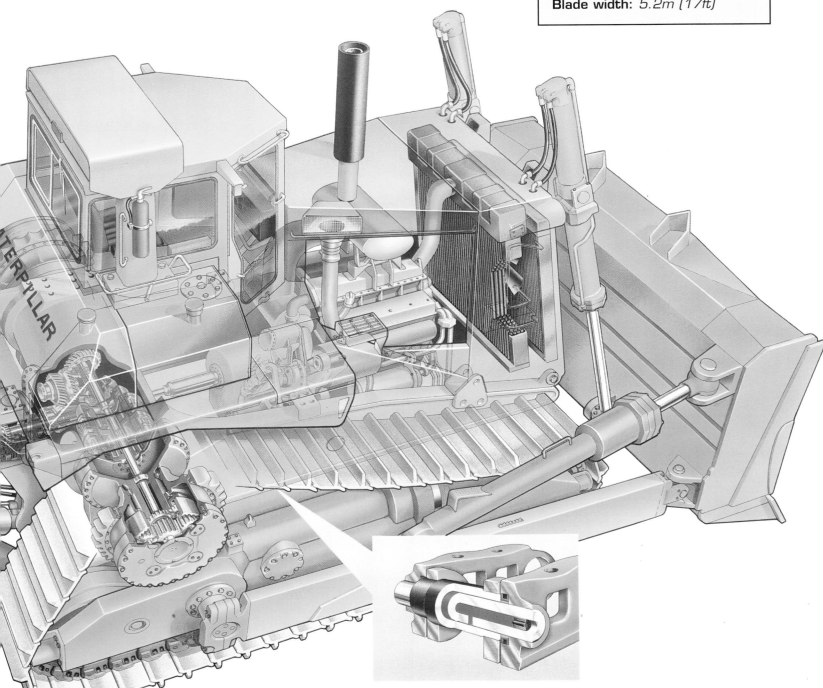

INTRODUCTION

Working against a backdrop of shuttering for massive concrete piles at a French construction site in the 1930s, a cable-operated dragline excavator loads wagons waiting on a railway line.

today's big names like Caterpillar and John Deere came to prominence in that period. The largest grader in the world is a Caterpillar 24H Mining Motor Grader, weighing 65 tons and powered by a 600bhp engine. As its name suggests it is active in minefield applications, and its rotating blade – or moldboard – is 7.31m (24ft) wide.

GRADERS

By their very configuration, scrapers are related to graders, and they developed along similar lines. One of the earliest was a mule-powered device designed in 1882 by James Porteous. Following hard on the heels of mule power came the steam traction engine, which could haul scrapers and graders with no problem, provided the inclines weren't steep. Next evolution came in 1915 with T.G. Schmeiser's aptly named Land Leveller, which was a scraper in which the blade was raised and lowered by compressed air. By now, scrapers and graders were hauled by tractors, and typically the early crawler models from Daniel Best and Benjamin Holt were prominent in off-road situations. The first all-welded, large-capacity scraper was the so-called Mountain Mover, built in 1922 by Robert G. LeTourneau, a legendary American innovator and heavy plant manufacturer. He went on to pioneer the use of rubber tyres and self-propulsion on his scrapers, and his site clearance equipment for Second World War and Vietnam War airstrips included the Tournapull Airborne Units and Tactical Crusher Units.

While LeTourneau paved the way, as it were, the post-war leader in scrapers was Butler Perryman, who owned the Gannon Manufacturing Company. His tractor-operated rollover scraper was acquired from the Earthcavator Company in 1954, and was superseded by the Hydraulic Landscraper model. Today, the far more sophisticated

States, and in the early 1900s the ground-breaking Holt two-ton crawler was fitted with a Russell Motor Patrol blade. By 1912 a tractor-hauled grader incorporating an adjustable blade mounted on a chassis with steel wheels was in production by the Galion Company, and this was soon in operation on road buildings and riverside levees. In 1919 the Russell Grader Company produced the first self-propelled grader, known as the Motor Hi-Way Patrol and based on a converted Allis Chalmers tractor. This reflected the pace of technological advances in engine manufacturing, although Russell continued to make horse-drawn graders alongside its tractor-drawn and self-propelled models into the 1920s. The company was bought by Caterpillar in 1928, and the first fruits of the amalgamation were evident in the Caterpillar Auto Patrol grader of 1931. The 1920s and 1930s were a time of intense road-building in the USA, and ever larger graders and scrapers were built to help with the interstate highway construction programme. The period was tempered by the swingeing cutbacks of the Depression, when a number of heavy-equipment manufacturers went to the wall, along with truck, automobile and motorcycle producers.

The Second World War had a double-edged effect on construction industry technology. On all sides, manpower and resources were directed to the war effort, detracting from relevant activities at home, while, as in any conflict, these technologies are honed to lethal perfection at a far quicker pace than in peacetime. As far as our subject is concerned, the most significant advance was in the incorporation of hydraulics, which benefited all types of earth movers from excavators to dump trucks. As the Western world reached new heights of prosperity, it was during the 1960s that earth movers had their widest applications, and several of

INTRODUCTION

devices from Caterpillar, John Deere, and Komatsu Dresser are the class of the field. A man could almost shave with one of these machines, so accurate are their scraper settings, having the advantage of laser-beam guidance systems.

When the bulldozers, graders and scrapers have done their work, a flat surface is sometimes desirable, particularly in road-building or construction sites, and this is where vehicles known as wheeled compactors come in. They are basically rollers of various kinds, normally diesel powered but originally steam driven. Some have a vibrating drum at the front and regular tractor-type wheels at the rear, while the more traditional kind have a single steel heavy roller at the front and a pair of steel rollers at the rear. So-called Padfoot drums have an embossed pattern on the front roller for use on sticky surfaces, and a further variant has a bulldozer-type blade on the front as well. When the compactor roller has performed its task, along come the concrete carriers, asphalt pavers and road rollers, which distribute the concrete or tarmac in the desired area, then, in the case of the asphalt, compact it to the required pressure. Giant paving machines such as those made by Gomaco lay concrete to create motorways and airport runways with widths of up to 45.7m (150ft) at a time. Other specialised versions are configured to lay tight radius curved pavements.

QUARRYING COLOSSUS

By far the largest heavy-duty plant machines are the monster excavators that work quarries and mines and carve out the landscape for road, rail and canal construction. They are the stripping shovels, walking draglines,

When the 134-m (440-ft) high Millennium Ferris Wheel was ready for erection on London's South Bank in October 1999, a special goat-leg crane was constructed for the operation. Although only a temporary structure, it was the world's largest crane, and raised the 1500 ton wheel with four hydraulic jacks pulling on cables attached to the wheel.

bucket excavators and tunnel-borers, and they devour the landscape for breakfast. They are so big and efficient, as well as costly, that they are something of a rarity, and they tend to stay in one place for a very long time. To make them cost efficient, some are shut off only for servicing, otherwise operating day and night non-stop.

STRIPPING SHOVELS

Stripping shovels look like giant excavators, and their buckets attack the hillsides with a forward motion – like a man with a spade – as opposed to the more versatile excavator that can operate with a scooping motion as well. They are stabilised by a complex system of lesser gantries and cables. Some of the big stripping shovels are still in operation in the USA, despite the fact that none have been built since 1971. The Marion 5900 was the last of this particular genre to be made, and that's appropriate, since the Marion company was first on the scene in 1911 with the initial stripping shovel, the 150-ton 2.7-cubic metre (3.5-cubic yard) capacity Model 250. Intended primarily for use in open-cast coalmines, the Marion Model 250 was quickly joined by the rival Bucyrus 175-B stripping shovel. Stripping shovels were mounted on a turntable base and moved on railway tracks that were extended as necessary, and in the late 1920s caterpillar tracks were used for motive power. In 1919, the Ward-Leonard control and drive system was introduced, and this has been a key facet of this type of machine ever since. The last stripping shovel to be built by Bucyrus Erie was made in 1965, the year that Marion came out with the world's largest stripping shovel, the 6360. Capacity of the 6360 bucket is 137.63 cubic metres (180 cubic yards), and the boom measures 65.53m (215ft) long while the whole unit weighs 12,631 tons. This unit is unique, and it is still in operation at Caption Mine in south-western Illinois.

A close relative of the stripping shovel is the walking dragline. These massive excavators sometimes have to be built on-site from components shipped in, and the prime example is Big Muskie, built in 1969 by Bucyrus Erie. This is the world's largest excavator, weighing in at 12,244 tons, and it operates at the Muskingum coal mine in Ohio, hence its name, where it can bite off 300 tons of soil with each bucketful. The principal is that the bucket is dragged along through the ground to be mined, while the rig itself moves at a snail's pace via two jointed shoes, each of which is 39.6m (130ft) long.

The first walking dragline was based on the walking mechanism patented by Oscar Martinson's Monighan Machine Company in 1913. This was followed the next year by the same company's

Mining and tunnel-boring machinery is extremely specialised. During construction of the Channel Tunnel, various types of equipment came into play, including the cutter-head borer, whose flailing heads (centre) eat away the rock face, and the full-face borer (right) that contains the whole tunnel infrastructure such as extractors, conveyors, cooling and props.

I-T walking dragline, which remained in production until 1925. Larger machines ensued as the walking mechanism was refined during the following decade. In 1939, the Lincoln-based Ruston-Bucyrus built its first walking dragline, the 5-W, which continued to be made until 1971.

Other walking dragline makers include Pages, whose 700 and 800 Series models have seen service all over the world, not to be confused with Marion's numerous 7000 and 8000 Series draglines, and Ransome and Rapier, which produced 56 models of walking dragline between 1939 and 1964. Production resumed in 1976 with the Rapier W2000, which had a bucket capacity of 34.4 cubic metres (45 cubic yards). In the former Soviet Union, the UZTM- and NKMZ-manufactured walking draglines were in use during the 1950s in Russia and the USSR and were exported around the world.

BUCKET-WHEEL EXCAVATORS

Possibly the largest and most complicated excavating machines are the huge bucket-wheel excavators that operate on open-cast coalmines. Some are built on multiple caterpillar-tracked chassis, while others move along a rail track. They consist of a system of gantries – usually four – with a central control module and a system of pulleys and cables that operate a large wheel mounted at the end of one of the gantries. This wheel will have between nine and 18 buckets attached to it, which dig up the commodity as the wheel rotates. Each bucket deposits its contents onto a conveyor belt within the gantry. The concept of the bucket-wheel is by no means new, having been used since ancient times to draw water from wells for irrigation, and Leonardo da Vinci is credited with the notion of using the bucket-wheel concept for excavation purposes in the fifteenth century. Even Leonardo would be impressed by today's bucket-wheel excavators, the biggest of which can move 244,672 cubic metres (320,000 cubic yards) of earth a day – enough to fill a football pitch to a depth of 180ft.

If the buckets are attached to the steel chain that rotates along the length of the boom, these machines are known as bucket-chain excavators. They originated in Europe in the late 1800s, initiated by Boulet of France and the German Lubecker concern, and were used specifically in mining and trench digging operations such as the Suez Canal. Significant producers of this equipment include the German companies O&K and Krupp & Buckau-Wolf, with machines capable of shifting 33,000 tons of material per day. Some of their machines are active in mine reclamation sites, filling and covering over the giant scars of worked-out mines and quarries with earth imported from elsewhere so the land can be used for agricultural purposes.

On a smaller scale, and closely related to wheeled loaders and tracked excavators, are trenchers. These vehicles have booms or blades that act on the same principal as a chainsaw and carve their way into the ground to create narrow trenches. Some can be up to 1.5m (5ft) wide and dig trenches 6.1m (20ft) deep.

INTRODUCTION

In Alaska from the late-nineteenth to the mid-twentieth century, bucket-chain excavators were used for gold mining. Positioned at the edge of rivers where gold was known to be, the prospectors' ramshackle clapper-board excavators – or dredges, as they were known – housed the sieving apparatus that yielded the precious metal from the sediment drawn up from the river bed by the bucket-chain boom. By the 1940s, they were finding some 22,680kg (50,000lb) of gold dust a year, but this environmentally dubious activity never recovered from a total shut down during the Second World War. Similar dredges have been used in other locations to extract tin – as in Indonesia in the 1970s.

While the massive excavators that work the open-cast mines on the surface can be as big as you like, those used in the mine-shafts below ground have to be rather more specialised to fit in the restricted galleries and coal-bearing strata, and they are often custom made for particular projects. They tend to be long and low, some operating on the bucket-wheel principal to rip away blocks of coal, while others use paired-up disc-shaped shears that scour the coal face in a ruthless manner. The raw material is removed via conveyor belts or low-bed shuttle cars, and a contemporary 1200bhp long-wall mining system using the shearing principal can produce 13.6 million kilograms (30 million pounds) of coal a day.

TUNNEL-BORERS

Probably the most specialised of the monster excavators are full-face tunnel-boring machines, which act like giant drills and perform their task using a rotating cutting head fitted with

The largest pieces of earth-moving plant in the world are bucket-wheel excavators. Looking like some kind of fairground ferris wheel, this Krupp model occupies the length of two football fields with its integral conveyor systems, and the 18 buckets rotating on its big ferris wheel churn away the over burden at an open-cast mine with ruthless efficiency.

INTRODUCTION

When the bulldozers, scrapers and graders have done their job the pavers and trimmers arrive on site to lay the finishing medium. This Gomaco 9500 trimmer is capable of laying aggregate to a specified computer controlled depth.

tungsten teeth. As they chew their way through the hillside, their progress is regulated by hydraulic push rams that move the rotating head forward. The attitude of the 'bit' is controlled by a hydraulic motor, which can adjust the cutting teeth forward or backwards, backed up by another electric motor that controls the rate of progress. The spoil is removed from the tunnel via a conveyor belt, and concrete liners are inserted into the tunnel walls as the borer advances. The whole operation is monitored by closed-circuit TV from a dedicated control room. One of the most recent tunnels to be dug in this way is Japan's 53.11-km (33-mile) long Seikan Tunnel, which was finished in 1985, while undoubtedly the most historically significant tunnel to be dug is the 50-km (31-mile) Channel Tunnel linking England with France, completed in 1993. The tunnel boring machines – or TBMs – used in this project were 15.25m (50ft) long and 8.84m (29ft) in diameter and weighed in at 1300 tons. In this case, the rotating head chopped away at the relatively slow rate of between one-and-a-half to three times a minute, chewing off 99.9 cubic metres (3531 cubic feet) of under-seabed strata every 1.5m (5ft). The other conventional tunnel excavating method is the boom-type cutter-head, which has the advantage of being much faster to set up and to operate, but in the final outcome is less efficient than the full-face method.

There are other even more specialised forms of heavy equipment that should be considered. The military has an insatiable appetite for many of the pieces of apparatus mentioned above, while farming too requires ever larger harvesting and planting equipment. Road-building is also specialised, demanding not only dump trucks and concrete carriers but machines for laying tarmac. The largest such paver and trimmer was made by Gomaco in 1988 to line the Californian Coachella Canal, and measured 31.4m (103ft) wide and 14.6m (48ft) at its base, specially shaped to accommodate the canal's sloping banks.

Now consider a rather different and often hostile environment. As you fly over the North Sea, you can look down and spot oil rigs far below. They may look tiny from the aircraft, but many of them are as big as modern office blocks, mounted on stilts or concrete pillars above the waves. They too are equipped with cranes and helicopter pads, and whole communities of perhaps 300 people live and work on them, drilling for oil sometimes 305m (1000ft) into the seabed below. One of the largest is the Gullfaks C, located just over 161km (100 miles) off the Norwegian coast, 29.9m (98ft) above the sea and towering to a height of 349m (1148ft). The land-based Texan and Middle Eastern oil fields present a different aspect altogether, being a forest of derricks reaching down some 7620m (25,000ft) into the oil-bearing strata.

The last item in the heavy-equipment arena is the crane, prominent in all cityscapes and breaking up most urban skylines. From docksides to construction sites, cranes of ever-greater complexity dominate the landscape. The tallest tower crane in the world is claimed by makers Kroll to be its hammerhead K10000, more than double the height of London's Tower Bridge, and capable of lifting 226,800kg (500,000lb) up to 121.9m (400ft) high with a reach of 91.4m (300ft). Predictably, it is expensive, costing 7.5 million dollars to buy one.

Heavy plant and equipment presents an array of machinery, often mind-boggling in its size and complexity and always impressive. As 'civilisation' expands, it's no coincidence that the rig that points to the distant future is equally awesome in its proportions and intricacy: the NASA Space Shuttle and its launching gantry are huge, and its dual crawler transporters 'Hans' and 'Franz' weigh a colossal 2.72 million kilograms (6 million pounds) each, providing a launch platform 79.25m (260ft) long and 70.1m (230ft) wide. The space shuttle goes at thousands of miles an hour, but the paradox is that the transporter travels at just 1.6km/h (1mile/h).

MACHO MEN

If the general impression has been created that these are macho machines, then the same is true to a greater or lesser extent of the people that operate them. The muck-shifting industry is a

INTRODUCTION

very small one in the grand scheme of things, according to my pal Gerald Stubbs, who worked on a variety of crawler dozers, scrapers, graders and excavators, from Senegal in West Africa, then France and the UK. Everyone is known, and folklore is endemic to it. It follows that a digger driver's reputation precedes him, and having a bad reputation doesn't get him much in the way of start.

There is no legislation covering training and certification, although some years ago, the insurance companies were becoming worried by the number of fatalities concerned with the operation of earth-moving machinery, and also by claims arising from damage by and to machinery. In order to formalise matters in the UK, the government set up a training centre at Bircham Newton in north Norfolk to train operators in the safe usage of plant machinery. This extended to a certification scheme whereby a driver's competence was judged by testers on particular machines. The upshot was that no contractor would employ anyone who didn't have the machines he wished to operate registered on his CITB (Construction Industry Training Board) certificate. Following on from that, the insurance companies won't consider claims now unless the employer can prove that the driver involved in an incident resulting in a claim has the relevant paperwork. Although earth-moving machines are fairly easy to operate, some people will never manage it, while the more adept soon get the hang of how to go about a job and how to put a machine to work. To be really good, a driver needs to have experience, and that takes time.

BUMPING AROUND

One rumour regarding the operation of earth-moving equipment long term is that it can cause kidney damage, due to the constant bumping up and down. Some States in the USA stipulate that a driver cannot spend more than a certain number of months on a scraper. But dump trucks are pretty comfy things nowadays, so there is no such problem there. Caterpillar scrapers have a swan-neck – the articulation between the tractor and the box – which has a nitrogen ram fitted into it that absorbs a great deal of the bounce in the machine. Early Cat machines, and other scraper manufacturers whose products use solid swan-necks, produce machines that are quite hard, and a driver could get bumped around quite viciously. Stubbsy believed that the cure for that lay in going easy on the right foot, but some drivers, like rally drivers, actually enjoyed getting all four wheels off the ground as often as possible.

Operating hours fluctuated, but they were generally based on seven-day week on-road contracts. Saturday was a half-day and Sunday was an eight-hour shift. The other five days were generally twelve hours, and sometimes longer. The operators used to have long weekends off now and again. Stubbsy recalled that on one contract he worked on at the Bristol West Dock, he and his colleagues were putting two 'ghosters' into a seven-day week. This

Deadweight compactors such as this Aveling-Barford road roller are essential for flattening the finish surface of the road or runway, which in this case is tarmac. The steel rollers can be ballasted for extra compression.

involved joining up two-day shifts by working through the night until they ran back into the dayshift. So in the course of the week they would only get five nights sleep out of seven. In some ways then, although the machinery had moved with the times, the operators' circumstances had not, despite the air-conditioning and computerised cossetings of modern cabs and cockpits.

As technology pushes back more and more barriers, the heavy equipment necessary to consolidate the new frontiers will become increasingly complex and high-tech. The biggest machines have tended to remain unique in their field, indicating that there are limits in certain areas – the walking draglines, for instance. But newer avenues have thrown up potentially larger and more sophisticated machines, and there will doubtless be others in the future.

EXCAVATORS

We rather take it for granted that there's always a piece of heavy-duty equipment available when something a bit more substantial than a hole in the garden needs digging. Nowadays, it's possible to rent a mini-digger from the local tool-hire depot to dig a ditch in your own backyard. However, from the pre-historic post-hole, to the hand-dug canals and railways of the eighteenth and nineteenth centuries, it was always simply a matter of hard graft with a spade and shovel.

Then in the mid-1800s, mechanical diggers which had been born of the Industrial Revolution started to make an impact. As these steam-driven earth-moving machines became increasingly more widely available, their principal advantage over the gangs of plentiful and relatively cheap labour became apparent: they could perform tasks far more swiftly. This didn't endear them to the navvies.

Largest and most bizarre of diggers is the bucket-wheel excavator. This top-capacity giant is made by Krupp GmbH of Lübeck, Germany, and weighs nearly 13,000 tons and moves around on 15 crawler tracks. Mounted on the rotating wheel, its 18 whirling buckets slice the earth at a rate of 191,000 cubic metres (250,000 cubic yards) a day.

One of the first recorded mechanical diggers was a steam-powered excavator which was mounted on a barge for dredging shipping lanes in shallow river estuaries. It was built on Tyneside, north-east England, by Grimshaws in 1796. It seems that this was an exception as, in Great Britain, most of the excavation and shoring-up work for the canal and railway network was done by hand by gangs of navvies (an abbreviation of navigators). The opposite was the case in the USA, where labour was neither especially plentiful nor readily available, and distances utterly vast by comparison. Here, it was more expedient to develop steam excavators to do the work, and that is why the first practical land-based steam excavator emerged in the USA. It was designed by William Smith Otis who, in 1834, commissioned the Philadelphia-based engineering company Eastwick & Harrison to build his Otis Steam Shovel. It was mounted on a chassis fitted with four standard-gauge wheel axles for operation on railroads. It consisted of a vertical steam boiler and a single, reversible steam engine mounted towards the rear of the chassis and operated by a fireman, who had a role similar to the fireman of a steam train. This powered the excavator arm as well as providing a limited means of propulsion when the machine was in operation. Up front, a triangulated wooden jib was suspended from a 3.05m (10ft) high cast-iron column, and this could be swung in a 90-degree arc to either side. The bucket-shovel pivoted from the jib, with the sophistication of a racking gear to set the angle and depth of its scoop. It performed the excavating task by means of a steam-driven chain pulley, rather like a manual shovel action. Once full of soil (or whatever it was digging), the bucket was tilted by the operator so that nothing fell out, and once the jib was positioned above the tip or wagon, the rear or underside trapdoor of the bucket was opened and the contents released.

The Otis steam shovel was first put to work on the construction of the Baltimore & Ohio Railroad in 1834, and two years later, it was patented as an American Steam Excavator. At least four steam shovels were made by Eastwick & Harrison to the Otis design, one of which was involved in the construction of the Boston & Albany Railroad in 1838. The second one was exported to England where it was used on the Eastern Counties Railway, and the other two were exported to Russia. Their inventor did not live long enough to witness much of his machines' success, as he was killed by one of them in 1842, at the age of just 26. His prototype survived him until it was broken up in 1905. The Otis line continued under the Otis-Chapman name after Otis' widow remarried, and subsequently as Chapman-Souther with the arrival of a new partner, finally disappearing around 1880. However, the Otis steam

The whole gamut of earth-moving equipment was brought to bear on beachheads such as this at Munda airfield, where US Seabees used excavators to create dispersal areas during the Second World War.

EXCAVATORS

This Krupp multi-bucket chain grader has a handling capacity of 600 tons per hour and a potential cutting depth of 43m (141ft). It can operate on gradients of up to 50 degrees, and is in action here at the Alsen Breitenburg chalk mine at Lägerdorf, north Germany.

shovel that came to England was rapidly copied by a number of British engineers who took advantage of the expiry of the Otis patent in 1860.

'STEAM NAVVIES'

Among the most notable was James Dunbar, whose 1874 steam shovel was designed along the lines of the Otis precedent and known as the Dunbar Steam Crane Navvy. What had been a common term for the predominantly Irish labourers who had built the canals and railways was transferred to their mechanical successors, thenceforth known as steam navvies, and this remained the case for nearly 100 years. And because they invariably ran on railway (railroad) tracks, they were described as Railroad Type Shovels.

Meanwhile, Dunbar's steam shovel was manufactured by the Lincoln-based steam engine manufacturers Ruston-Proctor & Company, and they immediately bought up all Dunbar's patents. Known as the Ruston-Dunbar Steam Shovel, the prototype was operational in 1875 with a firm of public works contractors called Lucas & Aired. It was the first of a production line running into hundreds, and many were involved in the construction of London's Albert docks between 1875 and 1880, and the Manchester Ship Canal from 1887, where no fewer than 70 Ruston-Dunbar excavators were employed. During the latter part of the nineteenth century, more docks were built at other ports and railway networks burgeoned. Roads were improved, municipal sewers and gas pipes laid, reservoirs dug, which were all accomplished by steam-driven excavators, supported by the armies of manual labourers.

While Ruston's steel-constructed two-cylinder device, with its origins in the largely wooden Otis steam shovel, had virtually cornered the market, the rival Leeds-based Whitaker Steam Shovel Company of Leeds dodged the patent issue by coming out with its turntable-layout 360-degree slewing system. This consisted of the excavator jib and bucket, plus steam engine and boiler mounted on a turntable aboard a railway truck. However, the principles established by the Ruston excavator of 1874 have never been superseded, apart from the ability to turn through 360 degrees and the incorporation of electric and diesel power sources.

The Ruston excavator weighed in at 32 tons, and featured a substantial triangular structure that support its jib of latticed steel, while its bucket arm was a steel-clad wooden beam. This mode of construction remained standard for many years. To adjust the depth of cut, the bucket arm could be extended or retracted manually by the cranesman by means of a large wheel. During digging operations, the steam shovel was stabilised by means of jacks mounted on outriggers. The early excavators used chains rather than wire cables because they could withstand the stresses of the bucket arm's slewing action better.

These Railroad Type Shovels soon began to grow larger, with horizontal boilers like those of steam locomotives. Weights of 90 tons became commonplace as more complex rolling stock was developed to carry them. Multiple engine set-ups were incorporated to perform the various different movements of the bucket-shovel.

THE NAME GAME

Unlike British digger manufacturers whose products were identified by the name of the maker, steam shovels in the USA were often called after the name of the town where they were made, partly in deference to the partnerships rather than individual companies that manufactured them. The two most significant US producers of excavators during the forthcoming century were Marion and Bucyrus, both named after the towns they were made in. The Marion Steam Shovel Company originated in 1884 in the town of Marion, Ohio, founded by a partnership consisting of steam-shovel operator Henry Bambart, agricultural engineer Edward Huber, and George King. They rejected the rail-track format, and instead their Marion Portable Steam Shovel was mounted on a steel chassis,

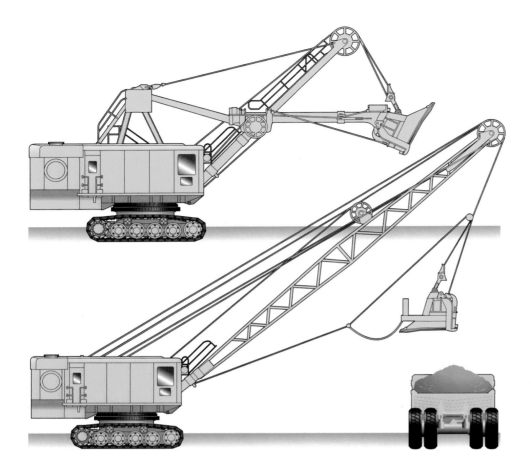

Draglines and electric mining shovels utilise an external power source, and both are cable-operated. They differ in their respective methods of grabbing and lifting earth, however. While a dragline draws its bucket towards itself via a long jib, the mining shovel digs and lifts its bucket with a boom and dipper handle.

running on wide-track, steel-treaded wheels, and the works consisted of a 360-degrees turntable. At the turn of the century, 24 of these Marion steam shovels were hard at work on the cuttings for the Panama Canal.

In 1911, Marion released a stripping shovel, designated the Marion Model 250, which quickly became popular in the US open-cast mineral mines, and at least one was exported to the ironstone quarries of the English Midlands. The difference between an excavator and a stripping shovel was principally in the longer jib of the steam stripping shovel. This was necessary, since its function was to clear away the overburden from above the coal or metal ore from the sides of the mine or quarry. The Marion 250 stripper operated on a rail track of sorts, a 3.6m (12ft) gauge and in short lengths that were moved around according to the area that was being worked. The jib or boom of the stripping shovel was possibly twice as long as a regular excavator because it had to access a greater area of quarry face, and in this respect, its boom was more like a crane's jib. The longest booms could measure 128m (420ft) long. A stripping shovel's bucket was known as a dipper, and was attached to a long arm called a knee or stick. Before long, the rail track was replaced by four separate crawler tracks that supported the machine and provided motive power.

THE CAPTAIN

For the next 50 years, Marion produced ever-larger stripping shovels, ranging from the 13.7 cubic metre (18 cubic yard) Model 5560 of 1932, to the 45.8 cubic metre (60 cubic yard) Model 5760 that appeared in 1956. Just under 10 years later, the Marion 6360 was launched, and that particular colossus weighed 12,600 tons, making it the largest electric stripping shovel ever built. The Model 6360 was known colloquially as The Captain in its workstation at the Captain Mine at Percy, Illinois. It cost its owners – the Southwestern Illinois Coal Corporation – 15 million dollars in 1965, and it was destined to remain unique. Its vital statistics were impressive, and indeed, it stands as the heaviest mobile land-based machine ever made. Its housing was the size of an office block, and the boom was 65.5m (215ft) long while the dipper stick measured 40.5m (133ft), enabling a cutting radius of 72m (236ft). Its movements were powered by 36 electric motors that were run off an external power source, generating between 200 and 460 horsepower each. As was the case with electric mining shovels, the electricity supply was extremely costly (perhaps 300 times that consumed by an average home) at a rate of 7000 volts, supplied via a 0.8km (0.5mile) long, 5cm (2in) cable. The Captain moved along on eight crawler tracks, each one 13.4m (44ft) long and 4.9m (16ft) high. The jib alone was 65.5m (215ft) long, while its four hoist cables were thicker than a man's wrist at 9.14cm (3.6in). The bucket was over 5.5m (18ft) wide, 4.8m (16ft) high and 7.3m (24ft) deep. All went well for 26 years. Then The Captain eventually had the misfortune to be destroyed by an accidental fire in 1991, by which time it had removed some 800 cubic metres (1046 cubic yards) of overburden.

The other principal manufacturer of excavators and stripping shovels was Bucyrus. The former maker of railroad rolling stock, Bucyrus was originally located in the town of the same name in northern Ohio, some 72km (45 miles) south of Lake Erie. The company came out with its first steam shovel in 1882, and it was involved in construction of the Baltimore & Ohio railroad. A decade later, Bucyrus was located in South Milwaukee, Wisconsin. The move hadn't diminished its output, which tallied 170 steam shovels in just 12 years. They were built under licence in Russia from 1900 and in Canada from 1903, and found work in opencast mines in the USA, as well as in the limestone quarries of the English Midlands, and in Spanish copper mines.

BUCYRUS-ERIE

The next phase in the history of the steam shovel was the acquisition by Bucyrus in 1927 of the USA's fourth major steam shovel manufacturer. This was the Pennsylvania-based Erie Steam Shovel Company, and the firm was accordingly renamed Bucyrus-Erie. By this time, a wide range of excavators and stripping shovels was being produced. The Bucyrus-Erie 3850-B had a 88-cubic metre (115-cubic yard) dipper bucket, and bigger

Electric mining shovels such as the P&H 4100 have sufficient muscle and dipper capacity to load a 240-ton dump truck in just three passes. The P&H 4100 has an 18m (60ft) boom length and a 45-cubic metre (59-cubic yard) bucket capacity.

still was its River King excavator, built in 1964, which worked the open-cast River King Mine at Marissa, Illinois. This monster weighed 8800 tons, and its bucket was a capacious 107 cubic metres (140 cubic yards), almost rivalling Marion's prodigious 6360 stripper. The River King worked full-time at the same location for 28 years until 1992, stripping the hillside for a grand total of 32.2km (20 miles), after which the mine ran out, and economics meant the monster's ignominious fate was to be reduced to scrap. This was in fact normal practice, since these behemoths were built on site from imported components, and were clearly too vast to transport to another mine. Another giant made by Bucyrus-Erie in 1963 was the 48.4m (160ft) high 1850-B electric mining shovel known as Big Brutus. It was erected on site at the Pittsburgh & Midway Coal Company's mines in Kansas, and worked pretty much non-stop for 11 years. It had a bucket capacity of 68.8 cubic metres (90 cubic yards), and during its lifespan, it uncovered over nine million tons of coal. However, it was the quality of the coal that spelled its demise, as it was high in sulphur and therefore environmentally unsound. Shut down in 1974, Big Brutus was revived as a museum piece in 1984 where its colossal size is a tourist draw.

Located across town from Bucyrus in Milwaukee was the third major US manufacturer of cable mining shovels, Pawling & Harnisch-feger, or P&H for short. P&H was founded in 1884 at the same time as Marion, although its first excavator, the P&H Model 210, didn't come out until 1924. A decade later, its first electric mining shovel, the 1.6-cubic metre (2-cubic yard) Model 1200WL was on offer. By 1969 its 19 cubic metre (25 cubic yard) 2800 Series mining shovel was in operation, featuring the P&H-developed Electro-torque system that converted AC to DC power.

EXCAVATORS

BUCYRUS-ERIE

Make: *Bucyrus-Erie*
Model: *395 B-III*
Manufactured: *1995*
Power: *AC ACUTROL GTO drive system*
Bucket-carrying capacity: *26.7 to 57 cubic metres (35 to 75 cubic yards)*
Boom length: *19.4m (64ft)*

One of the original manufacturers of face shovels and dragline excavators, Bucyrus-Erie introduced its 395-BIII electric loading shovel in 1995.

One of four 2800s was still in use at Elkwood Colliery, Sparwood, British Columbia in 1998, having logged some 130,000 hours of operational time. The P&H 2800 series remained in production, with capacity lifted to 35 cubic metres (46 cubic yards).

Going back in history, another English manufacturer of note was Priestman Brothers of Hull, founded in 1874 and specialising in steam-operated grabbing cranes and the grabbing buckets of the type that has a closing jaw, which could be fitted to almost any crane. Priestman's grabbing and dragline excavators were used extensively for dredging rivers and canals. One of the most popular was its Number 5 excavator, served up in 1924, and in production for the following decade. Revamped as the Priestman Cub, it could be used as a crane, a shovel, dragline or skimmer, and was still made in the 1960s.

Meanwhile, Rustons went into partnership with Hornsby in 1918, offering its highly successful No. 4 model a couple of years later. Because of the Depression at the end of the decade, Ruston-Hornsby struck up a merger with Bucyrus-Erie in 1930, becoming Ruston-Bucyrus. The Lincoln factory turned out bucket-type excavators as well as a small number of grabbing excavators. Most popular was the Ruston-Bucyrus No. 4, and no less than 934 units were produced between 1926 and 1933. This machine was available with several different types of power unit, including a twin-cylinder steam engine, a four-cylinder petrol or paraffin-fuelled motor, or a two-cylinder diesel engine. It was based on a sturdy crawler chassis with a 360-degrees turntable upper frame containing the controls and lifting and slewing gears.

DRAGLINE EXCAVATORS

The Ruston-Bucyrus No. 4 was of the excavator type known as a dragline and, being rather more akin to a crane with a bucket attached, its principal advantage over the excavator lay in its ability to dig or scoop material at a greater range than the regular jib and boom variety. The open end of the bucket faced the excavator and was attached to the jib by a cable or hawser. A second cable known as the dragline was attached to the front of the bucket to drag it back towards the vehicle, filling it up with the mud, silt, or whatever the contents were. The operators (there were usually two) then hoisted the bucket up and slewed the machine round on its turntable to offload the contents where required. This dragging motion of the bucket was the opposite to that of a regular excavator, where the bucket acted like a shovel, with the shovel arm pivoting on the jib in the same way as you would use a hand shovel. Draglines were more suited to dredging in harbours, reservoirs, and inland waterways, because they could perform their task at a lower level than their operating platform. The bucket was dropped into the water when the jib was at its fullest extent, at which point it would sink to the bottom of the waterway and the dredging process could begin. If the excavation was on dry land, the dragline worked its way along parallel cuts, uncovering the sought-after minerals and discarding the soil to one side of the cut. Having removed the mineral deposit, a fresh cut was started alongside the first one, and the overburden simply deposited in the empty trench alongside.

The Ruston Hornsby No. 4's main dragline steam engine was rated at 400hp and the slewing engines that turned it were of 200hp. Their coal bunkers contained four tons of coal, filled by means of a steam-powered hoist. They could also double as a crane if necessary, lifting 22 tons in a 38m (125ft) radius. By the 1950s, steam power had virtually been superseded by diesel or diesel-electric engines.

The dragline is said to have originated in Chicago, in 1903, invented by John Page, of Page & Schnabel, and positioned on a railroad wagon to participate in a canal-building project. It was Rustons who pioneered the dragline concept to an extent, and a measure of the machine's effectiveness can be gauged from an irrigation contract won by Rustons in Sikkur, India, in 1923, where some seven million acres was criss-crossed by a system of canals 64m (70yds) wide by 3.6m (12ft) deep. Ruston-Hornsby shipped out its machines knocked down in component form and its own engineers erected them on site. These 350-ton, steam-powered draglines had jibs 36.5m (120ft) long, while the dipper buckets could extract 10 tons of earth in a single cut. The operational cycle that included running out the bucket on the cable, dropping it, dragging it through the earth, lifting and dumping, was accomplished in less than a minute. Thus, each individual machine could fill 60 railway trucks an hour, the equivalent of some 191,138 cubic metres (250,000 cubic yards), which it was estimated would have taken 8000 manual labourers to achieve. That's how effective the machines were, and at least 12 were in operation on that particular contract.

WALKING DRAGLINES

It was normal for dragline excavators to move along on rails or crawler tracks, provided the terrain they were deployed on was firm. However, such was their weight that to be of any use when the going was soft, an alternative method of propulsion was needed. Meet the walking dragline. What sounds like it emerged from a sleazy nightclub actually originated in 1913 in the USA, courtesy of the Monighan Machine Corporation's Oscar J. Martinson. Its works – contained within the cab body, engine, latticework steel jibs, pulleys, counterweights and turntable – were low slung and based on a flat tub that gave it a secure base on the ground for going about the business of excavating. When it needed to be moved on a bit further, a pair of long, broad shoes located at either side of the machine were activated electrically by camshafts. The feet were connected to a large wheel on either side of the excavator. These revolved towards the front of the machine in an eccentric orbit, lifting the feet up and down. In the process the whole machine was lifted off the ground, shuffled along, and deposited a little further

This P&H walking dragline excavator is in operation at Bulga Coal Management's open-cast Bulga Mine at Mount Thorley, New South Wales. The machine 'walks' along by pivoting hydraulically on the 'feet' at either side.

EXCAVATORS

ACE OF SPADES

Make: *Pawling & Harnischfeger*
Model: *P&H 757 'Ace of Spades'*
Type: *walking dragline*
Boom length: *94.5m (310ft)*
Operating weight: *4417 tons*
Bucket-carrying capacity:
 *49.7 cubic metres
 (65 cubic yards)*

ACE OF SPADES

Ransomes & Rapier, originally a British firm, produced the W-1800 walking dragline, which had a bucket capacity of 25.2 cubic metres (33 cubic yards). One of these – fitted with a 30.5-cubic metre (40-cubic yard) bucket and excavating in an open-cast anthracite mine in South Wales – was claimed to be the largest walking dragline operating in Europe. This machine cost one million pounds when new in 1961. However, it was in action 24 hours a day, non-stop, and was finally dismantled and sold to a US mining company, having recouped its operator's original outlay.

By the 1990s, walking dragline manufacture was the province of British, Russian and US manufacturers, where 10,000-ton machines were not uncommon. The Marion Company, makers of the world's largest stripping shovel, also produced some of the biggest dragline excavators. The Marion 7400-M served at an iron-ore

The Land Rover defines the enormity of the P&H walking dragline aptly called the Ace of Spades, built in 1991 for British Coal's Stobswood Mine, Northumberland.

mine near Scunthorpe, Lincolnshire. However, Europe's champion dragline in the 1980s was known as Big Georgie, a Bucyrus-Erie model 1550, with a bucket sufficiently capacious to swallow a couple of cars. A P&H dragline, aptly named the Ace of Spades and ordered by British Coal for a cool 12 million pounds in 1989, was 18 months in the making, and finally went into action in December 1991. At the end of the decade, the Ace of Spades was still the largest such machine in Europe, in action around the clock – except for a half-hour break for its maintenance schedule – at the open-cast mine at Stobswood, Northumberland. Its projected lifespan was in excess of 30 years, by which time it ought to have written off its original costs.

forward. The pressure exerted by the shoes was considerably less than caterpillar tracks, so they didn't sink into soft ground. It was a bizarre procedure to observe. Walking draglines were operational in the London Brick Company's Ridgmont brickfields in Bedfordshire, England in the 1990s. Such was the size of dragline excavators that they tended not to travel vast distances. As a guide, however, back in 1974, a Ransomes & Rapier walking dragline covered 13.2km (13 miles) in the year.

EXCAVATORS

BIG MUSKIE

In the USA, the largest operational walking dragline at the turn of the twenty-first century was the Bucyrus-Erie 4250-W, which had 300-ton capacity bucket, or 168.2 cubic metres (220 cubic yards) of earth. The jib was 80.7m (265ft) long and the entire machine weighed a gargantuan 12,244 tons. Progress was understandably snail-like, at 0.1km/h (0.16miles/h). Since we're talking superlatives, the largest walking dragline ever was also a Bucyrus-Erie 4250 model, known as Big Muskie on account of its home at the Muskingham mine as the property of the Central Ohio Coal Company. Commissioned in 1966, it took three years to build the componentry – at a cost of 25 million dollars – and when completed, it stood 67.6m (222ft) tall and measured 148.4m (487ft) long from housing to the tip of the boom, and 46m (151ft) wide. Its bucket weighed 230 tons unladen, and had a capacity of 168.2 cubic metres (220 cubic yards). It was 4.2m (14ft) tall, 8.2m (27ft) wide, and 7m (23ft) deep. Like all walking draglines, Big Muskie relied on cables. In this case, they were 12.7cm (5in) thick, but its feet were hydraulically operated. On either side were 19.8m (65ft) long shoes, which lurched the big fellow forward by 4.5m (15ft) with each step. Theoretically, its pace was 274.3m (900ft) per hour. In 22 years, Big Muskie shifted some 465 million cubic metres (608 million cubic yards) of overburden, uncovering over 20 million tons of coal in the process. Then in 1991, the Clean Air Act prevailed upon the operator to cut back on production, and the monster was made redundant. Paradoxically, in spite of its sheer size, Big Muskie could be operated by just one person.

We've been mostly looking at leviathans until now. But at the other end of the size spectrum, and by far the most prolific excavators – in the UK at

Any doubts about whether size matters are dispelled by the prospect of Big Muskie, the largest ever walking dragline, in action at the Muskingham open-cast coal seams in Ohio.

BIG MUSKIE

Make: *Bucyrus-Erie*
Model: *4250-W*
Type: *walking dragline*
Power: *13,800 volts/62,900hp*
Bucket-carrying capacity: *288 cubic metres (220 cubic yards)*
Shoe length: *40m (130ft)*
Distance per step: *4.3m (14ft)*

least – are the bright yellow JCBs. The entire range operated via hydraulic-powered rams, and was available in a multitude of shapes and sizes, from mini-appliances just 1m (3ft 3in) wide that could dig a trench in your back garden, to more powerful backhoe wheeled loaders and crawler diggers. The backhoe method consists of the stick and boom of the front-shovel excavator, but orientated in the opposite way. It is usually seen only on the small- to medium-sized excavators. Instead of shovelling, the machine scoops the soil and draws its bucket back towards itself, when the operator slews it to offload. The configuration selected depends on operating circumstances, the job in hand, you might say. While a shovel excavator would be appropriate at the foot of a quarry, a backhoe would be appropriate on a cliff top, where it could eat away at the surface. Equally, the backhoe would be selected for any kind of trench-digging enterprise. Generally, front shovels were supplied with a general-purpose bucket, perhaps with cutting edge protection, or a bucket specifically for rock penetration, a high-density bucket, or a coal or lightweight spoil bucket. At the JCB level, bucket sizes are also interchangeable for different applications.

> **CAT 5080**
>
> **Make:** *Caterpillar*
> **Model:** *5080 front shovel*
> **Manufactured:** *1998*
> **Engine:** *14.6 litre Cat 3406C turbodiesel*
> **Power output:** *319bhp*
> **Bucket-carrying capacity:** *5.2 cubic metres (6.8 cubic yards)*

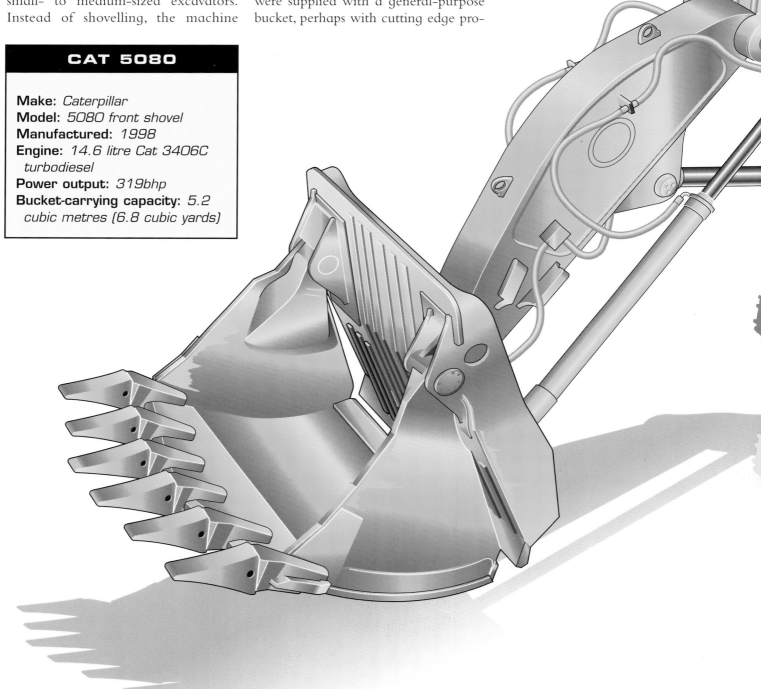

EXCAVATORS

HYDRAULIC ARMS

As mentioned elsewhere, hydraulic arms evolved tremendously during the Second World War, notably in aircraft controls, although hydraulically-operated excavators had their origins in the experiments of Sir W.G. Armstrong back in the 1880s. These involved subjecting fluids to pressure and transmitting power through pipes to activate rams, and several excavators with hydraulic functions came out in the wake of these tests. Post-war development – of hydraulic motors, pumps and, crucially, reinforced flexible piping that could withstand extremely high pressures – paved the way for the excavator models of the 1950s. Among the first manufacturers to dispense with winding drums and cables were Priestman Brothers in the UK and Koehring in the USA. The latter's hydraulic excavators featured booms that were longer than normal, and

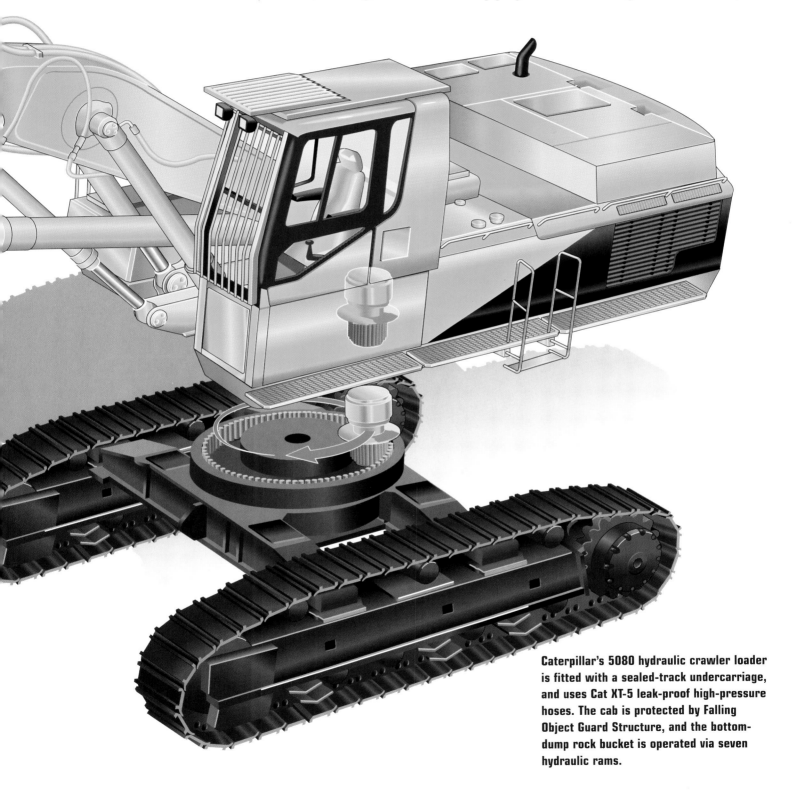

Caterpillar's 5080 hydraulic crawler loader is fitted with a sealed-track undercarriage, and uses Cat XT-5 leak-proof high-pressure hoses. The cab is protected by Falling Object Guard Structure, and the bottom-dump rock bucket is operated via seven hydraulic rams.

Left: Among the ubiquitous range of JCB hydraulic excavators available in the late 1990s was this 330LC tracked machine, operating with a backhoe configuration, the appropriate mode in this instance for loading a dumper from above.

Below: Krupp Födertechnik's gargantuan bucket-wheel excavator is the largest open-pit mining system in the world, and has been in action at Rheinbraun AG's Hambach site since 1978. Motive power comes from Siemens electric traction motors.

showed that a machine with a modest bucket capacity of about 2.3 cubic metres (3 cubic yards) was more precise than a cable version, whose cables were prone to stretch and sag. Another pioneer of hydraulic shovels and excavators was the French Poclain company, who originally used tyred-wheeled chassis. Poclain's Georges Bataille revealed his first design, the TU, at the Paris Agricultural Machine Show in 1950. By 1971, Poclain was offering the EC1000 model rated at 169.12 tons, subsequently uprated to 200.48 tons with the EC1000CK model.

ANY COLOUR YOU LIKE, AS LONG AS IT'S YELLOW

The Uttoxeter-based JCB concern was founded in 1945 by Joseph Cyril Bamford, who constructed a £45 farm trailer out of salvaged metal. Hydraulic tipping trailers were followed by hydraulic front loaders mounted on Fordson tractors. The company's first hydraulic excavator was revealed in 1954, designed around a Fordson Major tractor by JCB's Alec Kelly, and aimed at the agricultural market. One key feature was its 180-degree slewing arc, enabling a farmer to dig or dredge a ditch as the vehicle was driven alongside. After teething troubles to do with cavitation were surmounted, the Hydra-Digger was introduced in 1957, and in 1960 the archetype backhoe-bucket excavator was unveiled, designated the JCB 4. Unlike the five-lever control Hydra-digger, it had far more straightforward two-lever controls. By the mid-1990s, JCB was offering six mini-excavators from 1.4 up to 3.6 tons, and a pair of wheeled excavators in the 13- to 17-ton category, plus a line-up of 12 tracked excavators ranging from 7-tons to 45-tons capacity. There were a further eight models of the ubiquitous backhoe loaders, and as everyone knows, the JCB brand became the generic name for all such machines. Among the line-up was the JS220LC model with its 15m (49ft) long reach boom, ideal for implementing awkward jobs like clearing the banks of waterways or wherever access precluded close contact. The company's largest crawler excavator in the 1990s was the JS450LC, powered by a 200bhp six-cylinder Isuzu 6RBIT PE-01 turbodiesel with direct injection. JCB's CAPS (or Computer Aided Power Control System) continuously monitored its operating functions, which consisted of H for high production, S for general excavation such as trench digging, or L for precision of light-duty work. A backhoe model, the JS450LC could be specified with one of four different boom lengths and a choice of buckets.

Talking of longer arms, there was another manufacturer that specialised in long-reach excavators. This was the RB Variable Counterbalance Hydraulic Company, whose series of eight models combined the advantages of draglines and hydraulic backhoes. Its largest offering, the VC30, could reach up to 25m (82ft), its long arm balanced by a

EXCAVATORS

moving counterweight, and it came into its own in sand and gravel extraction, dyke excavation, or sea defence works. The RB VC20 was available with a deep-dig boom that provided twice the capability of regular booms.

THE CONVEYOR BUCKET SYSTEM

The endless bucket principle worked like a conveyor belt that had a number of buckets attached along its entire length, and the loop ran through a winch and along a boom. As it dipped into the quarry face or river bed, each bucket extracted a segment of spoil, which was carried along to the far end of its loop and then tipped out by gravity into a wagon or conveyor belt as the bucket went over the end of the loop and began its run along the boom again. Although the buckets were relatively small, there could be lots of them on one machine. The endless bucket system was first applied to dredgers clearing shipping channels of silt, and found its way into open-cast coal and mineral mining and marl pits for brickmaking. Originally steam-driven, the excavator itself was generally mounted on rails. By the 1920s, electric power

KRUPP BUCKET WHEEL

Make: *Krupp Födertechnik GmbH*
Model: *Giant bucket-wheel excavator*
Manufactured: *1978*
Crawler tracks: *15*
Power: *Siemens electric traction motors*
Buckets: *18*
Capacity: *191,000 cubic metres (250,000 cubic yards) per day*

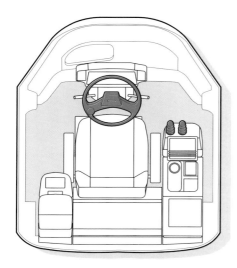
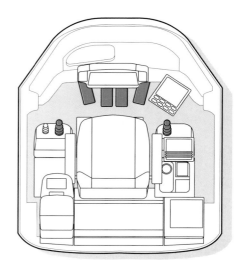

A hydraulic crawler excavator is relatively easy to operate as it has only two main controls, and two that operate the tracks. Two levers raise and lower the boom, push the dipper in and out, fill or dump the bucket, and slew left or right. Two floor pedals make it go forwards and reverse, with the direction-lever between.

was normal. The most significant producer of continuous excavators was the German company Buckau-Wolf and its evolutions included, in 1927, a slewing boom that supported the bucket chain. By 1939, it had come up with the biggest bucket-chain excavator in the world, with 38 buckets each with a capacity of 1514l (333 gallons). The entire device weighed 2500 tons.

There were one or two variations of the continuous bucket excavator. One was the ladder type that was largely the province of the German makers, which was in effect a dragline excavator running on rails or caterpillar tracks, with a second, longer boom fitted. Known as the ladder, this boom accommodated the bucket-loop conveyor belt, and the buckets on the underside were lowered on to the quarry face and dragged back towards the excavator, ripping away at the raw material in the process. The spoil was deposited when each bucket got to the end of the ladder to commence its outward journey again.

BUCKET-WHEEL EXCAVATORS

Yet another type of continuous bucket system was the wheel excavator, designed (on paper at least) by Leonardo da Vinci in the fifteenth century. Before the advent of steam power, they used wind or water or animal power. Early commercial uses were in the construction of the Suez Canal in the 1860s, and during the 1930s, they featured in German coalmines using electric power. These machines usually had 12 buckets mounted on the rim of a wheel at the end of an extended boom. The harder the surface, the more buckets were fitted. The boom was placed in contact with the surface to be mined and, as the wheel rotated, the buckets tore away at the coal or quarry face. The material thus removed was tipped immediately on to a conveyor belt to be hauled away for whatever the desired mineral was to be extracted. This was frequently coal, lignite or anthracite, limestone or bauxite. Bucket wheel excavators ranged in size from huge, 12,000-ton machines, right down to tiny, one-man operated vehicles. Historically, they were made by Ruston-Hornsby, Ruston-Bucyrus, Orenstein & Koppel (O&K), and Krupp and Siemens in Germany. One factor in their distribution was their incapacity to mine hard, rocky soils effectively, and they are more likely to be found in German brown coalmines and coaltips around the world. An example of one of the largest pieces of heavy plant ever made was the Orenstein & Koppel (O&K) 289 continuous bucket-wheel excavator. This machine weighed 14,028 tons and had a capacity of 249,490 cubic metres (314,000 cubic yards). The largest Krupp bucket-wheel excavator weighed just over 13 million kilograms (29 million pounds), and shuffled around on 15 crawlers. Its 18-hod bucket-wheel could scoop 191,138 cubic metres (250,000 cubic yards) of material every day.

A specialised branch of the excavator industry was the trencher, a device that could move on tracks or wheeled format like a JCB. One analogy would be to describe them as giant chainsaws, since their blade chains consisted of rasp-like claws, like crocodile's teeth, and the blade could be 6.1m (20ft) long and up to 1.5m (5ft) wide. The advantage it had over the mini-digger was that it removed only a minimum of spoil when performing its task. Certain alternative carbide-tipped blades could cut through rock or ice, and others could follow the cut by laying the cables. Some were fitted with a dozer blade to push aside the spoil. The manufacturer offering the widest variety of trenchers in the 1990s was Vermeer, followed by Ditch Witch and Trencore. Among the biggest was the Vermeer T-1455, weigh-

EXCAVATORS

ing 90 tons and measuring 18.3m (60ft) long, nearly half of which was blade.

Another type of continuous bucket excavator was the gold dredge. Half riverboat, half log cabin, these curious excavators served the remote gold mining communities in the Alaskan gold rushes that began at the turn of the twentieth century. By the 1920s, the most successful prospecting firm was the Fairbanks Exploration Company. Funded by speculators back on the eastern seaboard, Fairbanks built 10 gold dredges, consisting of a barge with a shabby wooden superstructure up to four storeys high, and this contained the ladder's winching mechanism, its filtration system of trommel barrel, gold table and riffle tanks, and spoil conveyor. Waste matter was dumped behind the dredge. The frozen ground could only be worked for perhaps 200 days of the year, and having thawed and removed much of the overburden with hoses, the machines went into action. The largest dredge could excavate 22 potentially gold-bearing bucketfuls per minute, and might travel as much as 1.6km (1 mile) in a season. Despite their primitive appearance, they were very lucrative and, in their heyday during the 1930s, they were producing 22,679kg (50,000lb) of gold dust a year. Largely abandoned during the Second World War, one of the Fairbanks dredges was opened as a museum in the 1990s.

MODERN TRACKED EXCAVATORS

An hydraulic excavator is composed of two main parts. The undercarriage consists of a track-frame, which carries the tracks, and between the tracks is the slew-ring, which is a large toothed ring. The tracks are driven by a final drive on either side, by hydraulic motors. The motors can drive either way, so each track may travel in either direction. The turret, for want of a better word, is the part of the excavator that revolves around the slew-ring. The turret comprises the engine and hydraulic pumps, the cab and the business part of the machine, which is in three parts.

The part hinged to the turret is the boom. At the end of the boom is hinged the dipper arm, at the end of which is hinged the bucket which has teeth, or not, as the case may be, depending on the application it's used for. Ditching buckets do not normally have teeth, whereas all excavating buckets do. The turret is driven around the slew-ring by an hydraulic motor, so that the machine can slew 360 degrees and another 360 degrees in the same direction without impediment. That's to say that the slew-ring has no stops fitted to it. The driver can slew back and forth as he wishes. The boom, dipper-arm and bucket are all hydraulically controlled by rams. The boom rams are

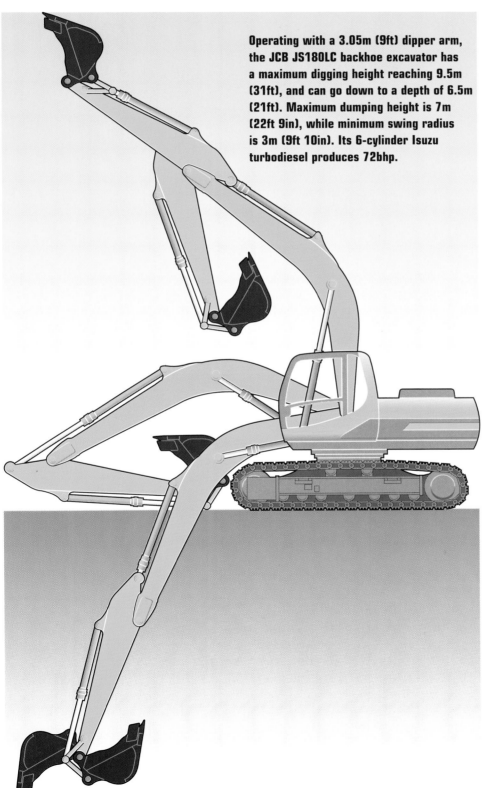

Operating with a 3.05m (9ft) dipper arm, the JCB JS180LC backhoe excavator has a maximum digging height reaching 9.5m (31ft), and can go down to a depth of 6.5m (21ft). Maximum dumping height is 7m (22ft 9in), while minimum swing radius is 3m (9ft 10in). Its 6-cylinder Isuzu turbodiesel produces 72bhp.

fitted to the turret at one end, and the boom at the other. The dipper rams to the boom, and the dipper and the bucket rams to the dipper and the bucket. When the machine is digging, the tracks are normally motionless, but the tracks are engaged to change the vehicle's position.

HAVE A GO

A hydraulic excavator should be an easy machine to drive as there are only two main controls, and a further two that operate the tracks. It is quite difficult to get the hang of, however, as the two main levers have two separate functions. Now you're going to get a lesson in how to operate it from Stubbsy. You're sitting in his Cat 235, while he stands on the running-board by the door of the cab and shows you how it works. On the floor are two pedals operated by your feet. One makes it go forward, and the other is reverse. You'll have to watch out which is forward and reverse, because as you're driving a machine whose turret may be facing backwards, you could set off backwards when you press the forward pedal. The thing to remember is that the drive motors on the tracks are at the back of the machine.

If you are digging away and you want to retreat a bit, relative to the way you're facing, just glance down at the tracks, and if you see the track-motors in front of you, then press the forward pedal for the machine to go backwards. In between the pedals is a lever. This is the sprag-lever, or direction-lever. When either one of the drive pedals is depressed, this lever becomes active. If the turret is the right way round, relative to the tracks, and you are moving forward, then if you move the lever to the left, the machine will go that way as the left track will stop while the right will continue to turn. If that doesn't give you enough turn, then you push the lever further, and the left track will reverse while the right one will continue forward.

Most machines now have a pedal for each track, and you can either toe or heel the pedal. Toe is forward on that track, while heel is reverse. This gives you total control of the tracks with your feet, and dispenses with the central lever. The advantage of not having to use a hand in the operation of the tracks is that you can keep them on the operating levers and use the bucket to assist the machine in getting out of awkward situations.

It's also useful for quite normal applications. For example, if you want your tracks to face 90 degrees to their present position, you can raise the machine off the ground, so that it is standing on the bucket at one end and the very rear part of the tracks at the other. You then work the opposite tracking on the pedals and turret-slew with the relevant lever in the correct direction, and though the turret will stay in the same position relative to the ground, the tracks beneath you will swivel around by 90 degrees! This is difficult, though not impossible, to carry out if you need a hand on a tracking lever. In this case, you use your knee.

The two main levers of the machine are set in the armrests of your seat. They control the four main functions of the turret, and there are two on one lever and two on the other. The functions are: boom raise and lower, dipper in and out, bucket crowd or dump, and slew left or right. The operation of a lever is set out in a 'plus' sign. Forward will operate one function, back will operate the same function but in the opposite sense. Likewise, left will engage another function while right will engage the same function but in the opposite sense.

The other lever behaves in the same way, except that it controls different functions. If, instead of following the plus sign to work individual functions, you go top left on the lever (imagine it's the end of one leg of a multiplication sign) then you will operate the two functions either side of that position. Thus, by feathering the lever between the two positive positions, you get the machine doing two things. If you do this with the two levers, it follows that the machine will be operating its four turret functions simultaneously.

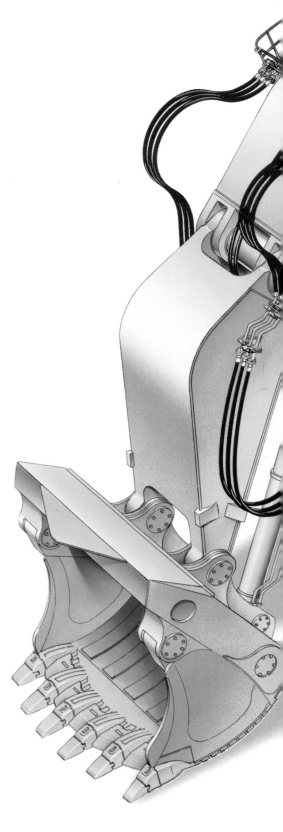

The internals of Liebherr's R984B unveiled, showing the pair of Cummins K1800E turbodiesels, crawler undercarriage with axial piston wheel motors, swing pump and swing motor for slewing, and the Litronic system that regulates the hydraulics that control the boom, dipper and digging bucket.

EXCAVATORS

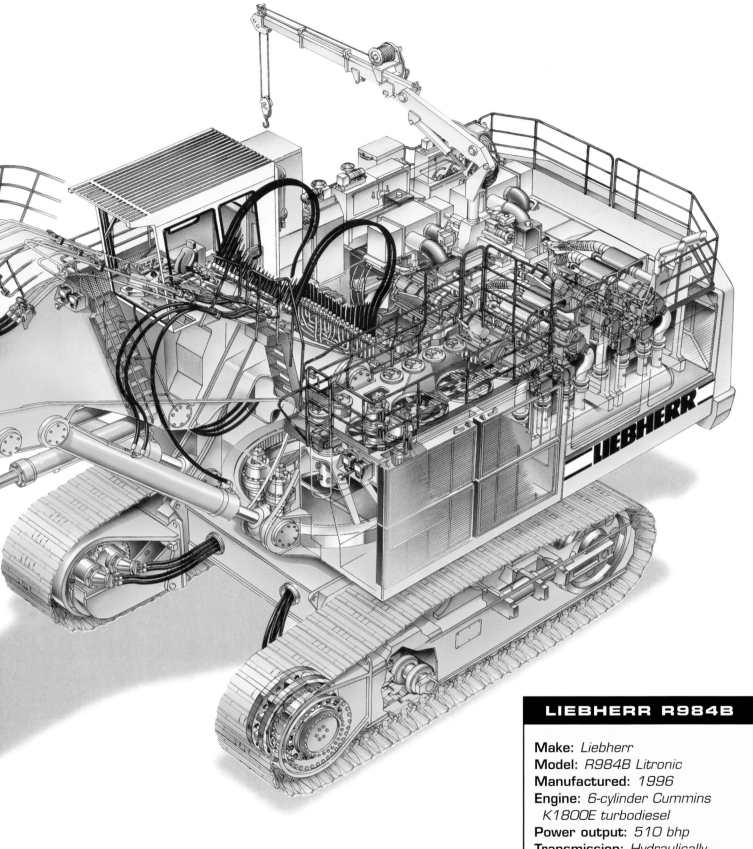

LIEBHERR R984B

Make: *Liebherr*
Model: *R984B Litronic*
Manufactured: *1996*
Engine: *6-cylinder Cummins K1800E turbodiesel*
Power output: *510 bhp*
Transmission: *Hydraulically-operated, variable-flow, swash-plate motors, via planetary reduction gears*
Bucket-carrying capacity: *33.9 cubic metres (44.4 cubic yards)*

Mining companies prefer to use equipment sourced from one company, and manufacturers' equipment ranges are structured accordingly. Liebherr, whose hydraulic excavator is busy here, bought the Wiseda mining hauler concern in 1995 to broaden its product base.

Stubbsy gives an example. 'You're sitting on the top of a dug face, with dump trucks below you, which you're loading one at a time. You've just got a bucket full of material. The bucket is clear of the dig and is still crowding, at the same time you're bringing the dipper in towards you, hoisting the boom and swinging round to where the dump truck is standing with its back to you. In this instant all four functions are engaged at the same time. There are even times when the four functions are engaged and the tracks are in operation as well!'

HYDRAULIC OR CABLE

The most common type of excavator in the 1990s was the tracked type. The small- to medium-sized ones were hydraulically operated, while the really big ones still used cables. The hydraulic type was sub-divided into front shovel and backhoe varieties, while cable-operated excavators came with shovels or dippers only. We're still in the realms of familiar names, including Marion, which was taken over by its old rival Bucyrus as recently as 1997. Prior to the buyout, Marion made 14 units of the 351-M, rated as one of its largest creations, which were exported to Australia, Canada, Russia as well as used in the USA. Its last two 182-M excavators were shipped to India. The 85-ton capacity Marion 351-M electric mining shovel superseded the 301-M model in 1995. These big machines retained cable methodology to operate them, rather than hydraulics. Finished in bright yellow, the first unit built worked at Suncor's tar sand quarry at Fort McMurray, in northern Alberta, while the third and last of the line was employed by the Fording River Coal Company at Elkford, British Columbia from 1996. This model in white and blue was fitted with a 43-cubic metre (56-cubic yard) shovel, or dipper, which also had a capacity of 85 tons. The operator's cabs were mounted on different sides of the body housing on these two vehicles, but the second one, which had a long jib at 22.8m (75ft) and a 64.2-cubic metre (84-cubic yard) shovel, had a cab on either side. It was in operation at the Thunder Basin Coal Company's Black Thunder Mine in the Powder River Basin near Wright, Wyoming, by 1996, and its white colour scheme contrasted with the commodity it dealt with. It also had the distinction of being the greatest capacity two-tracked coal shovel in the world.

When the Marion plant was closed down and its resources integrated with those of Bucyrus in Milwaukee, the 351-M was redesignated as a Bucyrus model, along with units of the Marion 8000 dragline series. Meanwhile, Bucyrus's electric mining shovels included the 495-BI model which, like the 351-M, had a capacity of 65 cubic metres (85 cubic yards). By 1999, the first Bucyrus 351-M had joined its elderly Marion sibling at the Suncor oil sand quarry in Alberta.

In 1991, that other maker of US giants, P&H, was offering its largest electric quarrying shovel, the 1225-ton 4100 series. With bucket capacities of 43–46 cubic metres (56–60 cubic yards), the 4100 and 4100A cable shovels were among the most popular excavators in the 85-ton capacity segment, with sales of more than 80 units during the decade. Among these was the 61-cubic metre (80-cubic yard) capacity 4100A LR coal shovel, in action at the Rochelle North Antelope Complex in the Powder River Basin, Wyoming. A 46-cubic metre (60-cubic yard) 4100 shovel could load a 240-ton capacity ultra-hauler dump truck in just three passes. These stripping shovels consumed 7200 volts of electricity at the rate of about 430,000kw over a 450-hour working month.

Other P&H 4100 excavators worked in the oil sand quarries of northern Alberta, designated TS for tar sand, and fitted with a 44.3-cubic metre (58-cubic yard) capacity bucket. Stability was enhanced by broader than average crawler tracks that were 3.5m (11ft 6in) wide. By 1999, three such machines were employed at the adjacent sites belonging to Suncor and Syncrude. P&H also manufactured the largest two-crawler mining shovels ever made. These were the P&H 5700 excavators, costing around nine million dollars each and launched in May 1978. Of the five units built, one suffered a dismal end. It was not so much an excavator as a dredger mounted on a barge, and was destined for the Great Lakes

Dredge & Dock Company. Named the Chicago, it was inundated by heavy seas off the coast of Denmark in 1996 while being towed to a fresh site, where it sank. Of the other four 5700 mining shovels, three were exported to Australia and the last one remained in the USA. In fact, this was the last P&H 5700 mining shovel made, delivered to the Coal and Allied Industries' Hunter Valley Mine in 1991. Like one of its brethren, based at R.W. Miller Pty's Mount Thorley Mine in Queensland, Australia, this unit was a 5700XPA model. They differed in having larger 44-cubic metre (57.5-cubic yard) buckets and operating weights of 2100 tons.

SHIFTING ALLEGIANCES

During the 1990s, large hydraulic excavators challenged the supremacy of the cable-operated mining shovel. The hydraulic front shovel enjoyed better mobility, was more selective about its target area, and could be positioned with greater accuracy. It could also be moved from one location to another more readily than the bigger cable shovels. However, while an hydraulic excavator might run to 60,000 operating hours, the cable shovel could easily surpass 100,000 hours and frequently outlast the seam it was mining. This led a number of manufacturers of the hydraulic models to seek mergers or joint ventures with other heavy plant manufacturers. Among them were some of the leading European manufacturers, including the long-established Orenstein & Koppel (O&K), Liebherr, and Mannesmann Demag. The latter came out with the first fully hydraulic excavator in 1954, but in 1995 it formed an alliance with the Japanese Komatsu company to produce hydraulic excavators under the brand-name Demag-Komatsu GmbH.

Japanese manufacturer Komatsu produces a wide range of hydraulic backhoe excavators, from the compact 10-ton PC95 to the 176-ton PC1600. However, these are relatively small fry compared with the 755-ton H685SP mining excavator with its 3730bhp twin CAT turbodiesels.

By 1999, Komatsu had struck a deal to take full ownership, and from that point, it was known as Komatsu Mining Germany GmbH. Largest of its offerings was the H655S front shovel, rated at 35.1 cubic metres (46 cubic yards), and the first unit to be built was dispatched over the ice-bound terrain to the Canadian Northwest Territories to Broken Hill Properties' Ekati diamond mines. A derivative of the H655S, known as the H740 OS, was built for Klemke Mining Corporation to dig in the oil sand quarries in Alberta. This unit had a wider excavator body and caterpillar track to provide greater stability during digging operations. It was powered by a pair of 1100bhp Caterpillar 3516B diesel engines and had an all-in weight of 815 tons.

By way of contrast, at the bottom end of Komatsu's eight-model range of light- to medium-duty tracked excavators was the 10-ton PC95 backhoe, powered by the 61bhp Perkins 100.4 diesel engine with hydrostatic transmission. The line-up featured Komatsu's patented hydraulic and electronic control system, known as HydrauMind.

In 1998, it was the turn of O&K to sell off its hydraulic mining equipment subsidiary to the Terex Corporation, and a new branch called the Terex

KOMATSU DEMAG

Make: *Komatsu Demag*
Model: *H685SP*
Manufactured: *1995*
Engine: *Two Caterpillar 3516DI-TA turbodiesels*
Power output: *3730bhp*
Bucket-carrying capacity: *35.1 cubic metres (46 cubic yards)*

EXCAVATORS

O&K RH400

Make: *Orenstein & Koppel*
Model: *O&K RH400*
Manufactured: *1997*
Engine: *Two Cummins QSK60 turbodiesels*
Power output: *4000rpm*
Bucket-carrying capacity: *42 cubic metres (55 cubic yards)*

Mining Division was duly formed, based at Tulsa, Oklahoma, along with its Unit Rig dump-truck wing. House livery was a uniform red, with white stripes and logo lettering. The O&K excavator package included three very big machines, identified as the RH170, the RH200, and the RH400.

GOLD MINE

Introduced in 1989, the 25.2-cubic metre (33-cubic yard) capacity O&K RH200 was the best-selling large hydraulic mining model over 500 tons during the 1990s, with 63 units sold worldwide. As an example of its diverse uses, RH200 number 46 was employed

in the mid-1990s at Barrick Goldstrike's gold mines at Elko, Nevada. A pair of Cummins KTA38C-1200 diesels with a combined output of 2102bhp powered the RH200. All-up weight was 529 tons. Its big sister, the O&K RH400, was the largest hydraulic excavator ever to be made. It was built at the company's Dortmund plant in 1997 after development in association with Syncrude Canada Ltd for use in its vast oil sand quarries at Fort McMurray in Alberta. These soils were particularly abrasive, requiring a particular specification of bucket jaw. Power came from two Cummins K2000E diesel engines, jointly rated at 3350bhp, although subsequent models enjoyed greater output of 3600bhp. Engine swaps in 1998 saw the new Cummins Quantum QSK60 diesels fitted in the original machine, which developed an astonishing 4000bhp. Bucket capacity of the RH400 was 42 cubic metres (55 cubic yards), which translated into an 80-ton payload. With an all-up weight of 910 tons, the O&K RH400 was some 155 tons heavier than the Komatsu Demag H485SP that previously held the robust record. From the ground to the top of the driver's cab, the RH400 was just under 9.2m (30ft) high, and the actual crawler tracks measured 9.9m (32ft 8in) long by 2m (6ft 7in) wide. The overall impression was of one chunky vehicle, and it presented a real threat in the traditionally secure market segment of the cable-operated excavators.

On a par with the O&K RH200 was Komatsu Demag's 25.2-cubic metre (33-cubic yard) H455S model that was released in mid-1995. The H455S weighed 540 tons and was powered by a pair of Cummins KTTA38-C diesels, booting out 2111bhp. The first unit to be built was employed at Suncor's Fort

Left: Excavators like O&K's RH400, the world's largest hydraulic front shovel, are fitted with buckets specially designed to cope with particular geological applications, such as the abrasive tar sands of Alberta, Canada. Power comes from Cummins Quantum QSK60 turbodiesels.

Right: Buckets are interchangeable according to application, and range from V-bottom frost bucket, top, with special teeth for shale or ice, lean-up bucket for easy clean-out, a chuck blade for ditch cleaning and back-filling, and the hydraulic twist-a-bucket that can tilt 45 degrees each way.

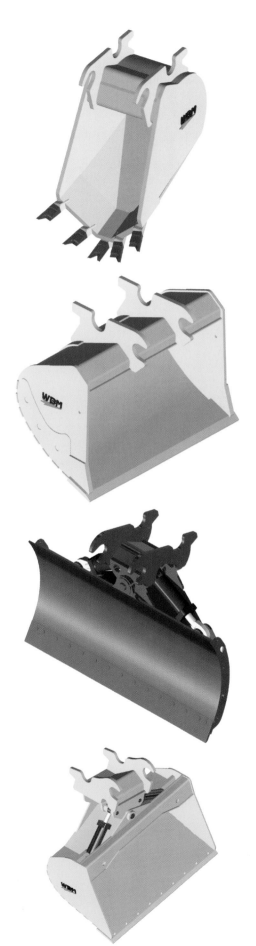

McMurray oil sand quarries in northern Alberta, having been featured at MINExpo in Las Vegas, Nevada in 1996. Rather bigger was the Komatsu Demag H485 SP, which had a 70-ton capacity rating and could accommodate 35.1 cubic metres (46 cubic yards) in its bucket. Its working weight was 755 tons, and its 3750bhp worth of motive power was provided by two Caterpillar 3516DI-TA diesel engines. Its typical environment was the oil sand mines of Alberta, in the service of Syncrude. This model was superseded by the H655S in 1996, and uprated features included redesigned front componentry, including the boom, stick, and bucket. It also was fitted with bigger and more powerful hydraulic tubes. Like its predecessor, the H655S was powered by a pair of Caterpillar 3516DI-TA diesel engines and, despite the revised shape, there was no actual increase in the bucket's 35.1-cubic metre (46-cubic yards) capacity.

BROADER PRODUCT BASE

The Franco-German Liebherr concern acquired WISEDA in 1995 from William Seldon Davis, awarding itself the prospect of a product base of rear-dump haulers and excavators. Prospective customers were more attracted by a manufacturer offering a broad range of equipment than one with perhaps only a single specialism. Built at Colmar in Alsace, eastern France, Liebherr's biggest face-shovel excavators in the late-1990s were the R992, R994, the R995, and the R996 Litronic. The latter machine was announced in 1995 and proved to be particularly popular in Australia. It was available both as a front shovel or backhoe format, plus a variety of buckets for different applications or material characteristics, and averaged 633 tons on the weighbridge. It was powered by a pair of water-cooled Cummins K1800E turbodiesels that between them developed 3000bhp, driving the four operating hydraulic pumps and the two slewing pumps. It could shift 33.9 cubic metres (44.4 cubic yards) with each bottom-dump bucketful, the equivalent of one 240-ton rigid dump truck in five passes. It carried 15,615l (3435 gallons) of fuel, and nearly half the R996 Litronic's bodyweight lay in its underpinnings, which consisted of three sections that could be dismantled for transportation. The slewing ring was an automatically lubricated triple roller design, and its tower and flange were made from a single casting. Its tracks consisted of pad-link crawlers with cast double-grouser pads. Each track was hydraulically tensioned and ran on three carrier rollers and seven track rollers, powered by hydraulic axial-piston wheel motors in consort with planetary reduction gears.

The slightly smaller Liebherr R995 was released in September 1998, and the first example was ordered as a barge-mounted dredger with a backhoe shovel. The regular P995 was fitted with a front shovel rated at 22.9 cubic metres (30 cubic yards), with an operational weight of 433 tons.

It would be extraordinary not to find Caterpillar present somewhere in this section about excavators. It appeared on the scene relatively late, and by the end of the 1990s, it offered a range of 40 tracked excavators. Smallest was the 55bhp Model 307, rising to the 428bhp 375L ME machine, and others were specifically long-reach, like the 325L LR, or demolition boom types, such as the 350L. The Cat 5130 hydraulic shovel or backhoe was designed to work in harmony with the Caterpillar 777C rigid dump hauler, rated at up to 110.23 tons. Customers were often more interested in buying into a package, and this was something Caterpillar excelled at providing. Its largest offering was the 351-ton Caterpillar 5230 model. This was launched in 1994 as an hydraulic front shovel with a 16.9-cubic metre (22.2-cubic yard) bucket, frequently seen loading haulers of up to 200-tons capacity. In backhoe mass-excavator format, the 5230 ME could be equipped with a 16-cubic metre (21-cubic yard) bucket for general duties, or a 27.5-cubic metre (36-cubic yard) bucket specifically for handling coal. The 5230 used a single Cat 3516 EUI diesel engine, churning out 1470bhp.

The Caterpillar 5130 was powered by the 608bhp Cat 3508 unit, which featured an economy device known as AESC or Automatic Engine Speed Control, and this cut engine speed from 1750rpm to 1350rpm whenever the hydraulic controls were out of action for four seconds. The cooling system minimised fuel consumption by means

of a hydraulic variable-speed fan operating at 400rpm. Like all such machines, the engine was housed within the box-section steel chassis frame, which was reinforced with braces at stress points, such as the end of the swing frame, boom and counterweight mounting points, and the final drive housing. Caterpillar's low maintenance crawler tracks were sealed and greased, featuring automatic track tensioning by means of a hydraulic gear pump to activate the idler. Top-class machines provided relatively comfortable operating stations for their drivers. Seats were multi-way adjustable, while the cab and surrounding componentry were mounted on viscous dampers to reduce extraneous noise and vibrations. A modicum of environmental control was available, like air-conditioning and sound-deadening. Lever and push-button controls were used along with joysticks and pedals, and systems were monitored via the VIMS or Vital

Like giant crabs on the beach, a cluster of Hitachi hydraulic backhoe excavators clears a rocky outcrop above a construction site. These are relatively small machines in the grand scheme of things, but are sufficiently agile to operate on terrain less accessible to bigger machines.

A matching pair of Cats at work. The 5230 pictured here, loading a 150-ton Cat 785B off-highway dumper, was the largest hydraulic excavator made by Caterpillar in the mid-1990s. It was equipped with a 17-cubic metre (22.2-cubic yard) bucket and powered by a 1450bhp Cat 3516 EUI turbodiesel engine.

Information Management System, with oil and fluid pressures, temperature, and engine speeds displayed. By virtue of its modular design, the 5130 could be broken down into eight segments for relocation.

FRUITS OF MERGER

A new manufacturer appeared at the end of 1993, born out of a merger between the Japanese Hitachi Construction Machinery Company and the Swedish company Volvo's north American subsidiary. It was based around Euclid's US operation, and appropriately enough, called Euclid-Hitachi Heavy Equipment, Inc. Ownership was split between Hitachi, with 80 per cent, and Volvo, with 20 per cent, and what was particularly unusual was that it was able to manufacture and market two distinct product lines from a single source. These were, of course, excavators and heavy-duty dump trucks. The excavator line included Hitachi's hydraulic mining shovels such as the EX2500 and the EX3500-3. Introduced in 1987, Hitachi's EX3500 hydraulic excavator was revamped in 1996 and designated the Super EX3500-3. Its dual Cummins KT38-C925 engines pushed out 1635bhp, and the 368-ton mining shovel supported a 18-cubic metre (23.5-cubic yard) bucket capacity. All things are relative and we're still in the big league here; but the rather smaller Hitachi EX2500 was unveiled in March 1996 at the CONEXPO mining show. This 14-cubic metre (18.3-cubic yard) loading shovel relied on a single 1254bhp 16-cylinder Cummins KTA50-C diesel engine. In backhoe format, its soil capacity was slightly less, rated at 13.8 cubic metres (18.1 cubic yards). As a front shovel, the machine's overall working weight was 264 tons, as against the 280 tons posted by the backhoe version.

Largest in the Hitachi range was the 570-ton Super EX5500. This model's bucket capacity was 27.1 cubic metres (35.5 cubic yards), equating it with the O&K RH200 and the Komatsu Demag H455S. It was unveiled at Syncrude's Aurora Mine, Alberta in July 1998 for operation by sub-contractors North American Construction. The Hitachi Super EX5500 was fitted with two Cummins KTA50-C diesels pushing out 2500bhp, and was at home in mining locations such as the vast Athabasca Oil Sands deposit in northern Alberta. These remote Canadian deposits were estimated to have several times more potential than the oil wells of Saudi Arabia. The oil was present in a primordial amalgam of sand, mineral clays, bitumen and water, and once mined, was hauled off for processing elsewhere. Resources such as these have been tapped for several decades using traditional draglines and bucket-wheel excavators. However, with hydraulic tracked machines increasing in stature and efficiency, the trend has been towards operators selecting the modern options, especially where a shovel capacity of 30.5 cubic metres (40 cubic yards) or less is required. Therefore it appears that excavators like the O&K RH400 and Komatsu Demag H655S are the way forward for heavy-duty mining operations.

You'll probably not come across such giants as these in your nearest uptown construction site. These are the province of smaller machines, produced by a wider number of manufacturers. Other players on the mining and quarrying stage, which also feature at the construction site level, include the Swedish Åkerman firm, part of the Volvo Construction Equipment empire and notable for producing small- to medium-size Volvo-powered hydraulic backhoe machines. Its largest offering in the late-1990s was the EC650 fitted with a 1.53-cubic metre (2-cubic yard) bucket.

At first sight, it's surprising to find Volvo offering a range of tracked excavators in competition with its own Åkerman subsidiary, which used the same engine line-up as well. But it happens within the motor industry all the time. For example, the VW Golf or Passat is mirrored by cars produced by Audi, Skoda, and SEAT, identical in all but body style and badging. And in the same way, excavator makers produce machines of comparable standards of service. Among the Volvo excavator range is the EC340, featuring an oil-bath slew system activated by an axial piston motor and planetary gearbox. Powerplant is the 190bhp Volvo TD 103 KAE that motivates the three hydraulic circuits which operate its boom and backhoe bucket systems.

Going out east to the Pacific Rim, the Korean manufacturer Daewoo Heavy Industries offered the 49-ton DH450 model at the top of a range of tracked excavators, powered by its own 221bhp turbodiesel V8 power unit. The Korean Halla Engineering and Heavy Industries relied on Cummins diesel engines to power its HE line-up of backhoe excavators, the biggest being the 40-ton HE 360LCH model. This used the Cummins LTA 10C, developing 190bhp, driving through an hydraulic two-speed transmission that powered it to a maximum speed of 4.76km/h (2.98 mile/h). Samsung was another Korean make, which came from the same stable as the shipping and heavy plant company. Its backhoe excavators were made at its Changwon Plant and marketed via its Illinois base in the USA. A typical offering in the tracked excavator department was its SE450LC-2, a 48-tonner fitted with the ubiquitous Cummins LTA 10C unit that served up 221bhp at 2000rpm.

The Japanese Kobelco brand was produced by the Engineering and Machinery Division of Kobe Steel, based in Tokyo. The range extended from the small 7.7-ton SK60 to the largest 50-ton SK460LC model. Most of these vehicles' functions were controlled by electronics, which in Kobelco's case, were identified as its Power Sensing System. The range-topping SK460LC Mk IV was powered by the 228bhp Cummins M11-C320 diesel engine.

It would seem that excavators of all types have a fairly secure future. So long as there is a requirement for the earth's naturally occurring minerals and fossil fuels, and as long as people want to construct roads and buildings, there will always be a role for the excavator.

Its bucket jaws full of rocky overburden, this P&H 1550 is operating at the Costain Mining Company's Pax coalmine in West Virginia. The 1550 was Harnischfeger's second-largest model when launched in 1989, and was powered by a single 12-cylinder Cummins KTTA38C-1350 turbodiesel.

MINING MACHINERY, TUNNEL BORERS AND DRILLING RIGS

We've already seen how vast and resourceful is the plant that's used in open-cast mining, consisting of huge face shovels, draglines and dump trucks. But underground mining is a different ball game altogether; only the commodity is the same. That's because the machinery's job is to claw large quantities of material from a seam face and extract it all from within the narrow confines of underground shafts.

To perform this task, a very specialised form of heavy plant has evolved. Its three main jobs are to score the embedded coal or mineral-bearing rock face to free-up the material, then extract it from the mine, while all the time supporting the earth lying above the mineshaft. No longer do mineworkers rely on the traditional pick and shovel to ply their trade. It's no less black and claustrophobic down a mine, but at least modern technology is capable of taking on what, 40 or 50 years ago, was one of life's grimmer career options. Now too, the tunnels are supported by hydraulic beams, rather than traditional wooden pit

Major tunnel-boring projects such as the one under the English Channel linking France and the UK are tackled by the appropriately-named full-face tunnel borers. This Robbins-TBM-529 operated by contractors Balfour Beatty was one of 11 machines used in the boring and construction of the 49.9-km (31-mile) tunnel under the seabed.

MINING, TUNNELLING AND DRILLING

Mining companies use special underground hauling equipment such as this Joy Mining Machinery 10SC27 shuttle car operating in the shafts of Coalberg No 9 mine. The shuttle car is less than 1m (3ft) high, but can carry over 17 tons of coal at just 8km/h (5miles/h).

props, which has a knock-on effect for the sawmill industry.

The other side of the coin is that the advent of new technology has improved productivity at the expense of jobs. In the US, the mining industry's workforce diminished by more than 20 per cent between 1987 and 1997, while productivity in coal mines actually rose by 30 per cent. This is all the more remarkable when you consider that some of the most easily accessible seams have been exhausted, requiring mining companies to access ever deeper and more secluded coal measures.

LONG-WALL MINING
Possibly the most significant advance in underground mining technology has been the technique known as long-wall mining. The long-wall mining systems are several hundred feet long, and consist of a series of powerful hydraulic jacks that support the roof of the shaft, while a slashing cutter-head tracks backwards and forwards across the coal face, dropping its produce onto a conveyor belt that carries it back up to the surface. A small team of miners operates the long-wall system, and at the cutting edge it consists of a pair of disc-shaped shears that are embossed with a variety of studs, shanks, picks and chisels that revolve at high speed to tear into the coal face. To get an idea of the principle, imagine holding an angle grinder with picks all round the edge of its disc, and pushing it head-first along a rock-face. The only difference is, the long-wall shears can be up to 3m (10ft) diameter, and one set can slice up to 1000 tons of material an hour. So we're talking boulder-size chunks of coal here, rather than lumps for your average fireside coal scuttle, and these, together with all the slivers, fall onto a conveyor belt running below the shears, to be transported to the surface. Meanwhile, the machine and its operators are protected from the weight of the earth above by stout hydraulic roof supports that self adjust to the height of the tunnel, which can be up to 6m (20ft) high. They also have the capability to shift along as the long-wall operation extends, and they can support some 7000psi (pounds per square inch).

As you'd expect with such a specialised activity, there are only a few manufacturers in the field, and among them are Joy Mining and Komatsu. The apparatus tends to be specially built to suit particular applications, and a typical order might call for 1000 roof supports, an armoured conveyor belt measuring 183m (200 yards) long, and a set of shears powered by a 1200bhp motor. As an indication of the capacity of such a system, the long-wall operation at the 304.8m (1000ft) deep Twentymile Coal Mine in Colorado chewed up a coal face 259m (840ft) wide by 2.4m (8ft) high and 5.6km (3.5 miles) long in just one month in 1995, yielding nearly one million tons of coal. That's some mining machine.

JOY OF MINING
Other types of underground mining equipment included the cutter-head type that was similar in concept to tunnel-boring equipment, although coming at the surface from a different approach. Their technique was to crack open-mine faces by means of immensely powerful cutter motors, with the operator shielded by overhead shuttering. Like the long-wall shears, this machinery excelled in cramped conditions, and the mining companies operating them were able to predict output and productivity.

Apart from the dedicated conveyor belts, mining companies used special underground hauling equipment in the labyrinthine tunnel systems of the mines. Among them were low, flat-profile shuttle cars made by Joy Mining that were under 0.9m (3ft) high and resembled a subterranean racing car, controlled by a driver who crouched in a niche at the rear. The difference here was that the shuttle car carried 18,144kg (40,000lb) of coal at no more than 8km/h (5miles/h).

Another denizen of the deep was the flexible conveyor train, made by Joy Mining and used for hauling smaller but nonetheless impressive quantities of

MINING, TUNNELLING AND DRILLING

coal. The 3FCT Continuous Haulage Flexible Conveyor Train, to give its full title, ran on wheels shod with rubber tyres, and could be specified as long as 152.4m (500ft) in length. Its capacity rating was 16 tons of coal per minute, travelling at a speed of 198.1m (650ft) per minute, and its serpentine versatility was such that it would also perform 90-degree turns where necessary.

Joy Mining Machinery is a subsidiary of Harnischfeger Industries, itself a Fortune 300 company, with global sales of equipment exceeding 2 billion dollars a year in 1999. With its origins going back to 1914, Joy has racked up over 75 years at the forefront of the development, manufacture, distribution and servicing of underground mining machinery for coal and mineral extraction. Among its catalogue of products, Joy Mining lists long-wall shears, hydraulic roof supports, face-to-surface conveyors, continuous mining machines, and batch haulage vehicles in the shape of shuttle cars and articulated haulers. In the late 1990s, it also made continuous haulage systems, meaning chain haulage and flexible conveyor trains, plus entry drivers, drills and loaders for use in long-wall and gallery- and pillar-mining applications. In 1999, Joy was the only manufacturer that offered a complete long-wall system to the mining industry.

CUTTER-HEAD TUNNEL BORERS

Back in ancient times, artisans made tunnels beneath pyramids, temples and city walls by means of mass slave labour equipped with nothing more sophisticated than hand-held chisels. Some 2500 years ago, Persian engineers dug an amazing 150,000 miles of tunnels for crop irrigation beneath the Iranian desert, which are still used as a water source by local farmers. Not until the Industrial Revolution did primitive hand tools start to be replaced by mechanical drills and dynamite for blasting.

The first significant developments in modern tunnel-boring machinery came out in the mid-nineteenth century, as a consequence of the burgeoning European waterways and embryonic rail networks which needed to tunnel through escarpments and hillsides that it would otherwise have been extremely costly to make cuttings through. Thus the first rock-drilling machines appeared around 1850, powered by steam engines and compressed air. One of the first applications was the 12.9-km (8-mile) long Mont Cenis tunnel below the Col-de-Frèjus, opened in 1871 and linking France and Italy at the point where Hannibal was meant to have crossed the Alps. In the USA, the 6.4-km (4-mile) Hoosac Tunnel in western Massachusetts was a significant precedent in mechanised tunnel boring. However, there was still a tendency to rely on the explosive nitro-glycerine, which although crude, did the business with the minimum amount of drilling, and its raw efficiency actually inhibited the development of rock drills.

Simple full-face tunnel borers were in use in Russian coalmines in the mid-1930s, although these were basically just enormous drills. You can create the same effect in miniature by fitting a router bit onto any household electric drill. Then in 1949 the Hungarian engineer Zoltan Ajtay came up with the prototype boom-system cutter and loader. Instead of going at the rock or coalface with a spiralled drill bit, the chain-operated cutter-head consisted of a series of rotating discs fitted with

A long-wall mining system consists of a series of hydraulic jacks supporting the roof of the shaft, and a cutter-head like this Joy Mining Machinery 4LS shearer, which tracks backwards and forwards at high speeds across the coal face, slicing off the raw material.

JOY MINING

Make: *Joy Mining Machinery*
Model: *4 LS long-wall shears*
Type: *continuous mining machine*
Manufactured: *1999*
Power: *1200bhp*

MINING, TUNNELLING AND DRILLING

Rock drills like this are used in surveying for exploratory probing to determine the geological composition of the tunnel walls, to prepare for blasting and to assist with stone fracture.

pick blades, attached to a crawler-mounted boom, and attacking the face full-on. As we've seen with coal-mining shears, the cutter-heads simply scarified a hole into the face, and the coal or mineral-bearing rock fell onto a conveyor belt, which was in itself a major advance. Previously it had been necessary to remove the drill in order to extricate the results of its efforts. A further refinement was the mounting of an engine in the boom itself, and this, coupled with refinements to the cutting discs it drove, led to major advances in tunnelling and mining efficiency. By 1960, an estimated 1000 cutter-head units manufactured by several specialist companies were operational in Soviet mines and tunnels and in other eastern European applications.

Cutter-head machines ranged in size from a 4.5m (15ft) to a 9.1m (30ft) boom. A typical cutter-head tunnel borer was operated by a single driver, and consisted of a crawler chassis, with a boom positioned at the front of the machine and a retractable shovel-shaped stabiliser blade underneath the front. The cutter-heads were cylindrical or pineapple-shaped, and could be equipped either with pick-like shanks or cutting discs. In a typical application, the cutter-head made an incision at the base of the rock face, into which was slotted the conveyor system. The operator then slewed the cutter-head up and down the face above that, and the spoil simply fell onto the belt and was carried away. While this was going on, the cutting crew erected the necessary steel shuttering that prefaced the lining of the tunnel walls. It was a much more direct means of tunnelling than a full-face system, and more convenient in that there was space for preparation work for the lining process to take place to the rear.

An interesting comparison between cutter-head machines and full-face borers was carried out by British Coal at Selby colliery, Yorkshire, in 1991. Using a set of each type of equipment, two tunnels were dug, and the full-face machine accomplished the given task at the rate of 100.6m (330ft) per week, which turned out to be twice that of the cutter-head machine. However, it had been possible to set up the cutter-head apparatus much faster than the full-face equipment, and the conclusion was that for smaller tunnelling jobs and forming galleries, the cutter-head option was the more flexible and economical. For anything longer than half a mile though, the full-face borer was the preferred option, particularly if the strata consisted of soft- and medium-density material. However, for especially stubborn rock such as quartz, there was still no substitute for the good old-fashioned drill and blast method, even in the 1990s.

FULL-FACE TUNNEL BORERS

Imagine a giant fan or propeller ripping into a rock face, and you've got the principal of a typical full-face tunnel borer. The business end consisted of between three and eight radial arms, equipped with discs and chisels, set in a cylindrical frame that could be from 1.8m (6ft) to 11m (36ft) in diameter. These arms rotated at speeds of between three to eight cycles per minute as they ripped into the rock face. The principal was that as the cutter spun the whirling steel discs and picks chiselled away the rock face, and the fragments tumbled into the bowels of the borer where a conveyor brought the debris away from the face, ultimately for dumping. The pressure that maintained the cutting edge against the rock face came from hydraulic rams that automatically braced themselves against the tunnel wall, and this could amount to a staggering 4.5 million kilograms (10 million pounds). Like underground mining equipment

MINING, TUNNELLING AND DRILLING

sets, full-face tunnel borers were made to special order, taking into account the nature of the rocks to be bored through, the overhead terrain, and length of tunnel. These factors determined the speed of rotation and the appropriate configuration of picks to be used. If conditions were likely to be wet, the equipment would be hermetically sealed and a special auger installed for extracting the core matter.

Apart from its relatively vast diameter, a full-face borer could be a lengthy beast too. Including its cutters and their driving mechanism, the conveyor that extracted the detritus, water pumps and tunnel-lining components, the full-face borer could trail over 305m (1000ft) back from the rock face. As demonstrated by the British Coal experiment, full-face borers were most cost-effective when employed on a long tunnel and drilling over a long period, due to high purchase and setting up costs. In addition, they worked best on softer geological formations, although toughened buttons and tungsten-carbide-tipped picks and chisels were employed for dealing with harder rocks.

One of the most recent tunnelling projects to come to fruition was the Channel Tunnel that finally linked the UK to the European mainland. It was by no means a new concept, and in fact the first full-face tunnel-boring machine was invented by J.D. Burton in 1875 with a view to fulfilling exactly that task. It was satisfactory as a basic soft-stone cutter, but simply wasn't up to the job of forging a 48.3-km (30-mile) tunnel. It was just 2.1m (7ft) in diameter, and incorporated a number of features seen on a modern tunnel borer. Up front, the cutter-head was attached to a sturdy iron frame that was braced against the sides of the tunnel. The cutter-head was made up of two face-plates, each of which bore six steel-disc cutters. It was clear from this configuration that Burton had correctly anticipated the two most significant factors governing tunnel-boring

Like evolved medieval weapons, cutter-head tunnel borers attack the rock face armed with swirling cylinders that gouge the surface to shards with a multitude of pick-like shanks. The operator moves the cutter arm up and down and from side to side to effect the onslaught.

MINING, TUNNELLING AND DRILLING

Business end of one of the Robbins-Markham full-face tunnel borers used in the construction of the Channel Tunnel. Assembled on site, the rotating cutting head consists of some 240 inset chisel bits that cut away the rock allowing the machine to burrow ever deeper.

equipment. The most efficient cutter-head is the disc method, while the forces generated require the machine to be rigidly braced during operations. There were mechanical deficiencies in Burton's equipment, and although other machines of similar design were tried subsequently, they couldn't sustain a concerted push through hard rock strata.

A US engineer called F.K. Mittry was credited with the design of a

Conveyor belt removes rock to trucks that take the rock away

Rotating cutter head

Cutter head is moved forward by hydraulic rams

Gripper pads anchor tunnel boring machine

FULL-FACE BORER

Make: *Robbins*
Model: *TBM-529*
Type: *full-face tunnel borer*
Manufactured *1974*
Engine: *7 x electro-hydraulic*
Power: *2300kw @ 11,000 volts*
Capacity: *100 tons per hour*

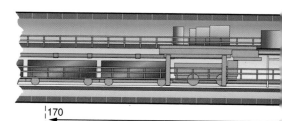

MINING, TUNNELLING AND DRILLING

7.9m (26ft) diameter tunnel borer, produced as a prototype by full-face mining manufacturers James S. Robbins & Associates in 1950. Designated the Model 910-101 Tunnel Boring Machine (or TBM), it was also known as the Mittry Mole, after its inventor. One of its first tasks was to burrow through the rocks of South Dakota during the creation of Oahe Dam. It had inner and outer cutting heads, which rotated counter to one another. The inner one had three cutting arms fitted with discs, the outer mustered six cutting arms, and the cutting heads were driven by a pair of 175bhp engines. A much more prosaic 25bhp engine was used to propel the 27.4m (90ft) long machine along its track. Its average speed through the soft shales was 49m (161ft) per day, which set a new tunnel-boring record.

A precedent had been set, and the way was open for more manufacturers to bring their own offerings to the construction industry market. In the late 1990s, full-face tunnel-boring machines were manufactured by Robbins, Lovat, National Tunnelling Equipment, Ebco, RML, Soltau, and Terratec Asia Pacific. In 1999, a Lovat RM129RL Series 9600 with a 3.25m (11ft) bore diameter would cost an operator $600,000.

THE CHANNEL TUNNEL

The concept of burrowing under the English Channel to link the British Isles with continental Europe was always a tempting one – for a potential invader that is, as on a clear day, you can see France from England, and vice versa. The 35.4km (22 miles) of sea that separated the two countries was tantalisingly narrow. Historically, from an English point of view, a tunnel was the last thing that was wanted if the country was to remain inviolate, and that stretch of open water certainly ensured that the UK wasn't physically invaded in two World Wars.

However, times change, and by the late 1980s, the funds and plans were in place to put the project into practice. Isolationists and anti-Europeans were outraged. But what was more surprising was that the Channel Tunnel was built in just over eight years. Not only was it the biggest tunnel in the world, it was also the largest construction project ever to be built by private funding.

The fact that most of the terrain under the English Channel (La Manche to the French) consists of chalk, which is a relatively easy substance to tunnel through, made the task easier. The political climate that evolved after Britain joined the EEC in 1970 led to an agreement in 1986 between the UK and France.

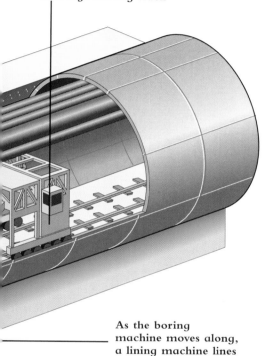

Part of train following the boring machine that lays railway track

As the boring machine moves along, a lining machine lines the tunnel with concrete segments

The Channel Tunnel is made up of three tunnels: two railway tunnels and one service tunnel that were excavated by tunnel boring machines. The cutting head is followed by a service train which copes with all aspects of tunnel construction from laying track and lining the tunnel to removing the cut rock and ventilation.

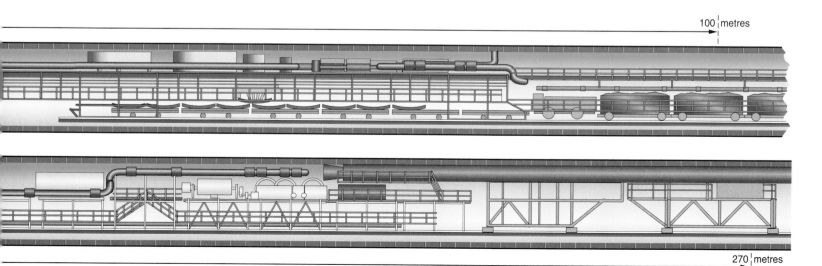

MINING, TUNNELLING AND DRILLING

Protracted negotiations between both governments and potential contractors resulted in a consortium of five British and five French companies known as Transmanche-Link being selected to construct the Channel Tunnel. The concession to own and operate it for 65 years was awarded to the Anglo-French company Eurotunnel. Work began the following year, with the French full-face tunnel borers beginning at Sangatte to the west of Calais, and British engineers starting off at Cheriton, near Folkestone. The fiscal tides ran this way and that as a succession of economic waves threatened to engulf the project. However, it was declared a going concern in 1994 when Queen Elizabeth II and President Mitterand opened it officially for commercial and passenger rail travel.

There were actually three parallel conduits constituting the 50-km (31-mile) Channel Tunnel. Two main tunnels, each 7.62m (25ft) in diameter, served the high-speed rail links north and south, separated by a smaller 4.9m (16ft) wide service tunnel. The three tunnels were linked by safety passages that intersected every 365m (1200ft), and a potentially lethal truck fire in 1996 proved their inestimable value.

CHUNNEL CHALK

In 1987, the US joint-venture company Robbins-Markham won a contract worth 15 million pounds to provide two full-face tunnel borers for the project, with designs for the back-up system and stress analysis provided by Bennett Associates. In the event, a total of

The Bucyrus-Erie 39R operating in this lunar landscape of a quarry is a rotary blast-drilling rig. Equipped with a three-headed conical bit for mining and mineral quarrying applications, it drills a number of holes in one site, into which explosive charges are laid. Compressed air delivered at 150,000psi clears the detritus after the blast, and then the excavators move in.

MINING, TUNNELLING AND DRILLING

This Bucyrus-Erie crawler rotary blast drill is operating in the floor of a quarry. Stabilisers spread the weight of the derrick, which is itself stabilised by a boom once erected by hydraulic rams. These big drills can penetrate to depths of 9m (30ft) and up to 56cm (22in) diameter. Speeds and pull-down pressures vary according to geological factors: iron ore is more than three times as dense as limestone.

eleven tunnel-boring machines were used on the construction of the 'Chunnel', with cutter-heads of up to 7.9m (26ft) in diameter. Different ones came into play depending on the geological conditions and moisture content of the chalk. Averaging just over three rotations a minute, with a thrust of up to 4220 tons, each water-cooled borer's eight-arm cutter-heads slashed the bedrock with their picks and discs, while the mashed chalky substance dropped onto conveyor belts running inside the body of the machine, handling 100 tons per hour. When the going was excessively wet, the borer body was sealed by watertight bulkheads, and an auger was used to extract the waste matter.

Like giant worms some 183m (600ft) long, the Channel Tunnel borers contained around 700 motors for a host of functions, from powering the blades, conveyor belt, cooling systems and hydraulic power that were fundamentally important, to driving the auxiliary pumps, fans and lighting generators. As each machine progressed forwards, the concrete rings that lined the tunnels were installed by construction engineers. Each one was composed of between five and eight segments, and measured 1.5m (5ft) wide.

The tally of equipment that was fitted prior to the tunnel coming into service was in itself mind-boggling. Much of the infrastructure of the Robbins tunnel borers was designed by Bennett Associates, who specified dimensions for the air and water pipes and high-pressure grouting system. More than 160km (100 miles) of railway lines were laid at an average depth of 45.1m (148ft), 20,000 lighting fixtures installed, 1200 telephones, 600 special doors and substantial ventilation, cooling, drainage, and, crucially, fire protection systems. The completed project came in twice over budget – a cool 15 billion dollars – having involved 15,000 workers, and extracted over eight million cubic metres (283 million cubic feet) of chalk, clay and soil. It's passed into folklore that at least one of the borers is still down there. When the French tunnelers were about to meet

MINING, TUNNELLING AND DRILLING

Some of the complex framework supporting a tunnel-boring operation is evident in this shot of a remote mine working. The mouth of the tunnel is at the left, with conveyors extracting earth and debris leading from it. Other conveyors and augers are linked to the quarry site.

the British crew in the middle, clearly the machines couldn't meet head-on. So the French one was driven off at a tangent and the final link made from the UK side. The French machine was buried in the side of the tunnel, while the British one was dismantled.

RIGGED RESULT

Tapping the earth's mineral resources is what this chapter's all about, and drilling for oil is no exception. The first prospector to successfully sink an oil well was Edwin L. Drake, in 1859. His Titusville-based West Pennsylvania well-head was pumped via a 6hp steam engine. Like any kind of prospecting, the notion of striking it rich provided a spur for the oil pioneers, with the result that rigs grew ever more sophisticated, and thrust ever deeper bore-holes into the earth. Prospecting companies ranged from Texan wildcat outfits to those of enigmatic Arab sheiks in the Middle East, with a host of maverick mobile operators active in the early 1900s. The demands of industrialised Europe and the USA spurred the production of oil as a lubricant in machinery. When production stabilised, the established oil fields of the USA and the Middle East formed the backbone of the world's oil resources. With the increasing sophistication of sysmological and geological surveys, it was possible to detect the black gold in increasingly remote strata. And with the growth and development of land-based oil-excavating equipment, it became feasible to source the product from secluded sites.

At the time that the Seneca Oil Company hired E.L. Drake to manage its surface oil well, the concept of extracting the viscous fluid by drilling was practically unknown. However, using equipment that had served in salt mines, Drake constructed the first oil-well. Flooding was countered by

MINING, TUNNELLING AND DRILLING

running the drill inside a cast-iron pipe, and eventually the oil started to gush when it was just 21m (69ft) below ground. It yielded between 10 and 20 barrels a day, which was twice the previous national product. By way of comparison, certain modern oil-well bores can be eight kilometres (five miles) deep. They work on a rotary principal, with an engine-driven turntable that spirals the drill bit into the subterranean oil-bearing strata. Offshore rigs employ the same method, with the proviso that the platform remains steady in spite of raging seas.

A new industry sprang up almost overnight with the discoveries of oil in offshore undersea locations, and while a desert-based oil field is impressive in its curious forest-like way, a towering oil platform far out to sea is a vastly more daring and spectacular proposition. Standing on massive jacks or concrete piers, these rigs can be higher than the Eiffel tower, and their construction has to be sturdy enough to withstand the rigours of turbulent wind, waves and weather. Where drilling sites are located in shallow water close to the shore, it's regular practice to build a pier out to the site for ease of access and stability. The maximum depth for an oil-drilling platform fixed by pilings into the seabed is around 30.4m (100ft). To go deeper requires the construction of a jack-up platform, which implies a rig built on a floating base that's towed to the targeted spot on the chart, where substantial jacks are extended from below the platform down to the ocean floor.

The South African Mossgas offshore oil and gas industry started in the 1960s when Soekor drilled the first offshore exploration wells with its drilling rig, the Sedko K. Testing carried out in 1980 in two areas known as the FA and EM fields revealed a potential harvest that would provide enough gas and oil condensate to produce 25,000 barrels of petrol and diesel every day for more than 25 years. The FA field was situated 85km (52 miles) south of Mossel Bay, with the EM field 49km (30 miles) to the west of the FA site. Exploration to establish the full extent of the petroleum gas deposits continued until the early 1990s. Mossgas Pty Ltd was established in 1989 after the South African government announced in 1987 that it would back a project to produce synthetic fuels from the FA and EM fields. The FA production platform towered 114m (374ft) above sea level and extended 105m (344ft) below it, making it one of the largest single structures ever made in South Africa. The feet of the rig, known as the jacket, were manufactured in Saldanha and were towed to the site on a purpose-built barge. In September 1990 the jacket was in its correct position and

Ready for action. A Robbins 167-267 full-face tunnel-boring machine comes on site, revealing the auxillary apparatus behind the cutting-head, amounting to a full tunnel-boring infrastructure, including water cooling, material extraction conveyors and prefabricated wall sections.

was anchored to the seabed. At this stage, a new spectacle in the form of the semi-submersible crane vessel, the Micoperi, came on the scene. This huge vessel, with a maximum displacement of 175,000 tons, had two lifting cranes each with a 7000-ton capacity.

ALL AT SEA

The largest marine oil-drilling rigs included the Gullfaks C, way out in the North Sea, which was 350m (1148ft) tall and resembled a village on stilts. Rigs of this type were based on giant concrete pillars, assembled by a complicated process in the water. The caissons, or hollow concrete supporting legs, were built in dry dock, and formed the base of the rig. The dry dock was flooded, the caissons floated out and were towed away to a sheltered site, at which point the supports were flooded and became submerged. Concrete pillars were constructed on top of the caissons, and the actual platform for the drilling apparatus, crew accommodation and service facilities like the helicopter pad, were built on top of the pillars. When all seemed ready to go, some of the water was pumped out of the caissons so that the rig bobbed up out of the sea. It was then hauled by tugboats to its work-station, and the caissons flooded again to sink the feet to the seabed.

The Gullfaks C rig was towed out to 174km (108 miles) off the coast of

Off-shore drilling rigs are constructed piecemeal in a complicated procedure and towed out to sea to the designated workstation. Like a small village, there's accommodation and facilities for the workforce, as well as the drilling apparatus, derricks and helipad. This is Shell's Nelson production platform, linked to the Safe Supporter Flotel and Rowan Gorilla II jack-up rig.

Norway, where it was positioned over existing wells to extract oil in a more economical fashion. The platform soared 29m (98ft) above sea level, and consisted of a helicopter pad, an enormous drilling derrick, five gantry cranes, and accommodation and facilities for the crew of 330.

MINING, TUNNELLING AND DRILLING

DRILLING RIGS

No less important are the numerous smaller-scale drilling rigs that are used in sourcing water, pile-driving in canal sides and breakwaters, and mining exploration. These rigs are a cross between crawler cranes and oil-field derricks, as many of them are just that – drilling derricks on crawler platforms. When I first moved to Norfolk and renovated a derelict farmhouse, the well that formed the water supply was polluted by nitrates. So a bore-hole had to be sunk, and the local drilling firm duly obliged. A truck-mounted derrick was set up, and a vertical power hammer pounded a succession of pipes into the field next to the house. As each section of pipe disappeared into the ground, a fresh piece was welded on, and the pipe was driven down until it met the chalk-bearing water-table. The artesian effect drew the water some distance up the pipe, and a pump at the top brought it all the way to the surface. The whole job took two weeks.

By the 1990s, a wide variety of truck- or crawler-mounted bore-hole drilling machines was on the market, and among the manufacturers was the German firm Krupp. Its product line included the DHR 80, which had double heads and was powered by a 73bhp Deutz diesel engine. Its pull-up capacity was 45,000N, its pull-down capacity 25,000N. The head rotated hydraulically, rated at 108rpm with a torque figure of 6200N, using 178mm (7in) pipes. The hammer was rated at 400Nm torque and the whole ensemble weighed 8000kg (17,637lb).

Liebherr's crane-type model HS 850 of 1982 had a 70-ton capacity, and could be equipped with a crane boom of 41m (135ft), which could be used for hammering or with a basic 13m (43ft) unit. The Liebherr weighed in at 46 tons without its 10-ton counterweight, and was available with additional equipment including hydraulic hammer and a Menck 15-ton air hammer.

The Bauer BG 9C that came out in 1992 was a hydraulic rotary piling rig, mounted on a Sennebogen S20 crawler carrier powered by a 166bhp Deutz diesel engine. Its torque rating was 7 tons and its drilling diameter was 1400mm (55in). Its crowd force was 150KN, and extraction force was 150KN. Total operating height was 14.6m (48ft), and it weighed 35 tons.

Poclain's 170 CK-B crawler-mounted piling rig appeared in 1991, also using a Deutz BF6L 513R diesel unit. Track width was 3m (10ft), and it was fitted with a Delmag leader 17.20m (56ft) long, and a Delmag hydraulic hammer, as well as two 5500kg (12,125lb) and 7500kg (16,535lb) winches. It tipped the scales at 36 tons.

The Ingersoll Rand drilling rig was powered by a 600bhp Cummins KTA 19C, with Cummins L10 secondary unit. It was equipped with a carousel for handling pipes, and had a mast capacity rating of 27,216kg (60,000lb). Pull-up was 13,608kg (30,000lb), and pull-back 11,340kg (25,000lb), while torque capacity was 3175kg (7000lb). The machine weighed 24,948kg (55,000lb), and was 10.7m (35ft) long, 2.5m (8ft) wide, and 3.9m (13ft) high.

The Klemm KR 901 hydraulic crawler rig came with a hydraulic hammer, clamps and a dust collector. Ideal for quarrying applications, it was powered by the 75bhp Deutz engine and weighed 6500kg (14,323lb).

The Massarenti MR 5000 of 1986 was a trailer-mounted rig, with a hook load capability of 113 tons. Its tower measured 34m (112ft), and it used an Ideco rotary table. The Massarenti was equipped with three engines. Main unit was a Detroit Diesel 16V 92T, with a Detroit Diesel 8V 92TA back-up, and a mud pump driven by a Gardner Denver PZ8. Tool range included hydraulic shaker, de-silter, and de-sander. Service rig derricks like these came in several configurations, including what were known as stiff singles, telescoping doubles, slant rigs and horizontal masts.

If this all sounds a bit idiosyncratic, consider that all trades and industries have their jargon and designations for specialised products and applications. Fortunately, the basic function of these machines is simple enough to grasp, and after all, we'd be stuck without them.

Celebrations are in order as the Robbins full-face tunnel-boring machine achieves a breakthrough at the Muela Dam site, part of the Lesotho Highlands water project in South Africa. The cutting heads of these machines can be configured with blades appropriate to the particular nature of the local geology.

BULLDOZERS

The business of moving earth around must be deeply entrenched in our psyche. Take a small child to the beach, and it starts work on a construction site of hugely ambitious proportions. The first toy I remember being given was a Dinky Toy's Blaw-Knox bulldozer. It was red and had a driver and rubber crawler tracks, and its blade appeared to work on a hydraulic principle, just like the real thing. Unlike the real thing, which runs on steel crawlers, its rubber tracks proved to be its Achilles heel, because they weren't robust enough to withstand the rigours of the childhood sandpit. It was a sturdy model in every other respect, and outlived many of its successors.

That, in a way, sums up bulldozers. They are mighty strong beasts, and form a vital part of the arsenal employed by contractors at quarries, mines and construction sites all over the world, used in conjunction with scrapers, excavators, wheeled loaders and dump trucks. Of the 115,000 or so crawler dozers operational in the USA, the majority are in use at construction sites, and the remainder at mines and quarries.

There are two basic types of bulldozer, and they are generally defined as crawler dozers, which have caterpillar

At the heart of any major construction project you can expect to find the bulldozer, also known simply as a tractor, a blade and a crawler dozer. Although a blade like this Cat D9 achieves its goals through brute power, its operation requires a good deal of skill.

tracks, and wheeled dozers which, rather obviously, run on wheels. The 'bull' part of the name has rather lapsed in recent times, hence the use of the word 'dozer' on its own, certainly in US parlance. In fact, as far as people in the UK muck-shifting business are concerned, the machine is known simply as a 'blade', and never called a bulldozer, although in the construction industry, it is sometimes referred to as a 'tractor'.

Like contemporary tractors, blades were originally built with conventional gearboxes and clutches. The engine transmitted its power to the gearbox by means of a main clutch, which was operated by a lever to the left of the driver's seat. The operating technique was explained to me by my old pal, Gerald Stubbs, who operated all sorts of construction equipment in Senegal, France and the UK. They were mostly Caterpillar machines, but the basic functions for most hydraulic plant are comparable. The lever was spring-

International Harvester's BTD-6 was introduced in 1953 and built under licence in the UK, shown here clearing muddy topsoil off a greenfield site. Originally offered as a tracked loader with a bucket, the blade was a later fitment.

loaded so that it stayed where it was put, either engaged or not. In between those two positions, the driver had progressive engagement. The gearbox drove into a differential, which transmitted the drive out to the left and right of the machine. Each half-shaft was fitted with a multi-plate clutch and a band brake.

Two levers on the centre console operated the clutches, and two pedals on the floor operated the respective brakes. The forward/reverse lever was on the right-hand side, and the throttle lever was positioned to the right of the clutch levers on the console. In order to turn the machine, the driver first disengaged the clutch on the side that would be the inside of the turning circle, and then applied the brake on the same side. This had the effect of putting all the power out on to the outside track, and braking the inside track that had been deprived of power.

With these conventional gearbox machines, there was no need to move the throttle lever from the full-throttle position in order to change gear, or go forward or reverse, as the main clutch lever also operated a brake on the gearbox main-shaft. At any gear change, the machine would stop, so the driver only

selected a gear for the work in hand. Gerald gives an example of how the blade worked in operation. 'You would be working in second gear, for instance, and at the end of the push you would declutch, change to fourth, and select reverse with the forward/reverse lever. Then, back at the start of the piece of work, you would do the opposite and get back into second and select forward.' Sounds easy? I think not.

POWER SHIFT

In the 1960s, everything went power-shift, although the manually operated blades as described above were still available. With a power-shift dozer the main clutch is eliminated as all the gears are via a torque converter, so it was impossible to stall the engine during sensible use. The first power-shift dozers still had the track clutches and brakes set up as before, but the difference lay in the gear operation and the throttle control. The gears were controlled by a single lever situated in a little console by the driver's left hand, and there were three gears forward and three reverse. In fact there were always as many gears forward as reverse because, by the nature of its work, the machine spends as much time in reverse

BULLDOZERS

Like this 8-cylinder 745bhp Komatsu D475A, all crawler dozers change direction by means of track clutches and brakes, stopping drive to one set of caterpillar tracks or the other, depending on the desired direction of travel.

as in forward. The slot the lever worked in was 'U' shaped, the bottom of the 'U' being to the front of the tractor. The leg next to the operator was 1,2,4, forward, the other being 1,2,3, reverse, while the bottom of the 'U' was where neutral was. When a change of direction from forward to reverse was made, the revs of the engine had to be cut, because they used a torque converter that drove an epicyclic gearbox. This was done by means of a decelerator, a small pedal like the accelerator in a car which was operated by the right foot, and which, when pressed, cut the engine revs. This obviated the task of pushing the hand-throttle in and out, as the driver simply left it in the wide-open position and controlled the revs with the decelerator.

The next development in dozer- or blade control was to put the track clutches on to the brake pedals. As the driver pressed the pedal, the clutch disengaged, and a further movement operated the brake. The clutch levers on the console remained in place and could also be used for other functions.

The tracks on a crawler dozer or blade are driven by sprockets. The vehicle's half-shafts drive exposed sprockets, and the shafts normally go through a reduction gear before meeting the sprockets. The sprockets engage the track chain – which is just that, a massive chain – on which are mounted the cleats or pads. The entire track is a wear-part, just the same as a tyre, and the grousers – the parts that stick out of the cleat to provide grip – wear away. The chains themselves also wear out, as do the sprockets. In 1977, Caterpillar realised that the sprockets, working at ground level, were subject to maximum wear. They consequently changed the design on the models D10 and above, lifting the sprockets off the ground, so that by the time any particular portion of track had got up to the sprocket, all the abrasive soil would have fallen off it.

In the cabs of these tractors, a radical change took place during the 1970s. The combined clutch and brake levers were shifted over to the left-hand side, and the operator's seat was set at a fairly acute angle to the right, so that he had a good view down the left side of the machine, the side he worked from.

NO DOZING OFF

So you want to drive a bulldozer? You'd better climb up on this Caterpillar D9G – a classic crawler dozer if ever there was one. Stubbsy says it's the best tractor that Cat ever made. This one is fitted with a spring-blade, and you're going to be pushing scrapers in a cut.

You've just pushed a scraper through the cut and have gone from third gear forward straight into second gear reverse, with a dip on the decelerator with your right foot as you changed through the 'box. The tractor gains speed and you push it into third, accompanied by another dip of the throttle. Looking over your shoulder, you see a scraper grading into the cut on a trajectory that will take him just past the left-hand side of your track, which would be appropriate. You have to slew round behind him, straighten up, and then go forward and pick him up to push him through the cut.

This manoeuvre is accomplished in the following way, which is the quickest way possible. As soon as his rear tyre has passed the middle of the tractor, you give it full-left slew by pulling in the left clutch lever with your right hand and applying the left foot brake. When the middle of the tractor is lying along the longitudinal axis of the scraper, you go from third reverse to second forward – with a dip on the decelerator – while keeping the brake on and the clutch disengaged. This has the effect of bringing the business end of the tractor in line with the scraper and, as soon as that's achieved, you let go the clutch and brake and catch up with the scraper, changing up into third if necessary.

You drop your blade on the approach to the scraper's pushing button, and just as you're about to make contact, you decelerate a bit. The top of the blade will make contact with the button, and as the blade is hinged at the bottom, you'll see that you've made contact as the top will come towards you. At this point you raise the blade, so that the button is nestling in the concave curve of the blade, and you let go the decelerator, having changed to second. At the same time, you put the blade into float. The blade won't drop to the ground as it's held by the button and the force of the tractor pushing it against the button. As the button is in the hollow, the blade cannot drop.

If your colleague on the scraper makes too heavy work of the job, by being enthusiastic on his dig lever, you may have to drop into first. You do this without dipping the decelerator, because if you did you might lose contact. When the scraper is charged and starts coming out of the ground and going up through the gears, you go up into third. And before you lose contact with him, you snap your blade into hold. The reason that you had the blade in float is because the relative motions of the scraper and the pusher are different, and if the blade was in hold there would be a conflict between the blade and the button. When it's in float it follows the motion of the button. As soon as contact is broken with the scraper, go into reverse and scramble back to the beginning of the cut.

This time, the scraper has finished grading-in, has set his cutting-edge into the ground and has stopped, waiting for you. The same procedure applies, although it's more precise. As soon as you pass his rear wheel, you go through the same procedure as before, but you'll endeavour to swing the blade within a few inches of his button so that when you straighten up and go forward you're straight on it and pushing – assuming that you managed to avoid clipping his rear tyre as you came round. This implies performing several operations in a very short space of time to get it right.

After you've pushed this scraper, you get back to find there aren't any others waiting, so you park your blade at 45 degrees to the cut where you want the next one to start. When he comes in, he will pass in front of your blade, and you go forward until you are on his longitudinal axis, at which point you line him up and button-up. Simple, isn't it ?

It might come as no surprise to learn that the noise of your Cat D9G is deafening. The exhaust is only four or five feet away from you, and only 20 or 30 years ago it was un-silenced. The exhaust came straight off a turbocharger that emitted its own very high-pitched whine. The clatter of the crawler tracks is secondary to the noise of the engine and transmission, although the noise of the tracks is audi-

Bulldozers

ble at least a mile away from a working machine. No wonder veteran crawler dozer drivers are a little hard of hearing.

BACK TO THE ROOTS

The crawler dozer or blade has its roots in the products of the Caterpillar company, dating back to 1925 when the rival Holt Tractor company and C.L. Best Tractor company joined forces and formed the Caterpillar Tractor Company. The first use of crawler tracks is attributed to the Holt Brothers' No. 77 steam-powered tractor, which featured a set of tracks instead of rear wheels. Subsequent Holt traction engine prototypes used tracks, and No. 111 was the first sold in 1906. Two years later Holt went over to petrol engines, and during the First World War, much of the company's production was delivered to the US Army. Its chief competitor, C.L. Best, supplied the farming community. Litigation was not infrequent over use of the 'caterpillar' trademark, and other firms active during the second two decades of the century, including Bates, Cletrac, and Monarch, were obliged to pay Holts a licence fee for manufacturing track-laying vehicles. One of the sub-contractors who had built the earth-moving plant for Holt and Best was Robert Gilmour LeTourneau. In 1928 he devised the Power Control Unit (or PCU), a tractor-

KOMATSU D375A-2

1: Komatsu SA6D170E turbodiesel engine
2: 22-cubic metre (28.8-cubic yard) capacity blade
3: hydraulic arm for blade tilt
4: single shank ripper
5: torque converter
6: idlers
7: sprockets
8: chassis frame
9: track frame
10: blade-tilt hydraulic piping
11: cylinder block & piston
12: halogen working lights
13: driver's cockpit
14: roll-over protection system
15: computerised monitoring system
16: radiator guard
17: air-conditioned cab
18: tinted glass
19: hydraulic system demand valve
20: control levers & pedals
21: ripper shank point

Cutaway illustration showing the vitals of Komatsu's D375A-2, a series launched in 1987 and progressively updated. Power comes from a 525bhp Komatsu SA6D170E turbodiesel, and the machine has a blade capacity of 22 cubic metres (28.8 cubic yards).

BULLDOZERS

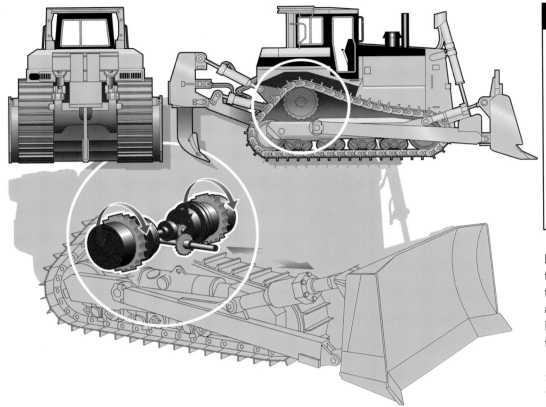

CATERPILLAR D10

Make: *Caterpillar*
Model: *D10N*
Manufactured: *1987*
Engine: *12-cylinder Cat 3412 turbodiesel*
Power: *520bhp*
Blade capacity: *21.9 cubic metres (28.7 cubic yards)*
Blade measurements: *5.9m x 2m (17ft 3in x 7ft)*

From 1977 onwards, Caterpillar shifted the main drive sprockets on its D10 model to an elevated position, so the tracks adopted a triangular configuration. This kept the vulnerable sprocket teeth away from abrasive rocks.

mounted, engine-driven, heavy-duty cable system that became standard on his equipment. By 1934, LeTourneau was making bulldozer blades that utilised the PCU. These were supplied to Caterpillar for its newly introduced diesel-powered Model Diesel Sixty, which, like all Caterpillar products of the time, was supplied simply as a bare tractor unit. It was only around this time that commercial vehicle producers were starting to offer diesel engines in road-going trucks. LeTourneau set up agreements with Caterpillar dealerships to stock his range of tractor accessories, which included not only the big bulldozer and angle-dozer blades, but scrapers, wagons, tree-rooters, and the PCU set-up. LeTourneau's range of products was sufficiently well matched to Caterpillar tractors that rival blades were gradually ousted from the market.

CONSTRUCTION ARENA

During the mid-1930s a transition took place which saw Caterpillar tractors move away from the agricultural sector and into the construction arena. Both companies prospered. During the Second World War, Caterpillar and LeTourneau products were in huge demand, and were heavily involved in ground clearance operations for the Allied military, particularly in construction of wartime airfields in the UK and invasion beach-heads in Europe and the Pacific. The principal weapon here was the Caterpillar crawler tractor allied to a Carryall scraper, with a LeTourneau bulldozer blade leading the way.

Despite the success of the combination, in 1944 Caterpillar decided to start making its own attachments for bulldozers, scrapers and controls to go on its crawler tractors. LeTourneau was divorced accordingly. Caterpillar's D8 series emerged as the key earth-moving vehicle in post-war construction sites, having been developed from the Diesel Seventy-five which had come out back in 1933. This was succeeded by the RD-8 in 1935 and the renowned D8 model a couple of years later.

In the post-war boom, other manufacturers appeared on the scene to challenge the Cat D8's supremacy. One of these was Allis-Chalmers, which released its 163bhp HD-19H crawler dozer in 1947. The same year, International Harvester came out with its powerful 180bhp TD-24 bulldozer, which remained the most powerful machine available until 1954, when Allis-Chalmers brought out its 204bhp HD-21 model. But waiting in the wings were two even more dynamic machines. These were the Caterpillar D9D series, and the Twin-Power Euclid TC-12 from General Motors' construction industry subsidiary. Caterpillar's prototype D9X was first shown in 1954, with the pre-production D9D coming out the following year. It's fair to say that probably no other machine had such an impact as the D9D. Powered by the 286bhp Cat D353 diesel engine, it had an operational weight of 32.3 tons, and from its launch in 1955, it was regarded as the benchmark by which all other bulldozers were judged. It was likely to be fitted with a Balderson U-shape blade, which was available up to 7.3m (24ft) in width. Power output rose to 320bhp in 1958, climbing to 335bhp the next year. Clearly, these ratings were important, because the more powerful the engine, the more earth it could displace. A further upgrade in 1961 to the popular D9G series set the standard for the next decade.

The Euclid Twin-Power TC-12 was announced in 1955, and was up and running as a prototype by then. It stands as the first production superdozer, by virtue of its twin-engine format. The Euclid TC-12 was fitted with a matching pair of six-cylinder Detroit Diesel 6-71 engines that developed 365bhp, matched with twin Allison Torqmatic transmissions, and this made it the most powerful and manoeuvrable crawler bulldozer of the period. Its distinctive appearance was due to its rounded bonnet contours, the driver exposed in a central seating position, and round-topped, rear-mounted radiators. It came in Euclid's distinctive lime-green paint finish, which made a change from Caterpillar's ubiquitous yellow. The next evolution of the TC-12 developed 413bhp, and by 1958 the TC-12-2 model was available, pushing out 425bhp with a displacement of 42.8 tons. If the Euclid TC-12 had a downside, it was simply that it was more complicated and more expensive to maintain than its rival, the Cat D9D. Ten years after its original launch, the TC-12 was given a revised designation of 82-80 BA, although the basic specification was still the same. In 1967 the model was upgraded to the 82-80 DA, which turned out to be the most powerful Euclid model, rated at 440bhp. When GM relinquished the Euclid name and trademark for its bulldozers and dump-truck lines in July 1968, the 82-80 was marketed under the Terex banner. When GM decided to discontinue manufacturing the 82-80 in 1974, production totalled 901 units, including all TC-12 and 82-80 models. By this time, it had been surpassed in the dozing stakes by the Allis-Chalmers HD-41, unveiled in 1970. It was a much more angular machine, featuring a driver's cab, which was now normal fitment, and it had the distinction of being the first bulldozer to crack the 500bhp barrier. It had its origins in a prototype built in 1955 using a twin-engined set-up, via another dual-engine prototype shown at the 1963 Chicago Road Show. The prototype was redesigned that year following reliability problems, and fitted with just a single 500bhp 12-cylinder Cummins diesel engine. During that whole decade, efforts were made to refine the design through extensive trials, and seven further prototypes were made. By 1970, the final configuration was arrived at, and the HD-41 was powered by the 524bhp 12-cylinder Cummins VT-1710-C diesel engine. All-up weight was 64.2-tons, and its 6.1m- (20ft-) wide U-shaped blade could move 21.4 cubic metres (28 cubic yards) of earth and rocks.

In 1974, Fiat's construction equipment division amalgamated with Allis-Chalmers and accordingly, the HD-41 became the Fiat-Allis 41-B, although in European markets it retained its original designation. Indeed, power ratings and specification remained the same, although its appearance changed with the adoption of the roll-over protection system (ROPS) cab. That meant that it had a roll-over hoop and rock-shield. By 1982, the 41-B had been uprated to the Fiat-Allis FD50 model, which was capable of sustaining a regular 525bhp and hitting 550bhp at high revs. It was powered by the Cummins VT-1710 diesel unit, and weighed in at 72 tons. Sometimes accused of deficient final drive reliability and electrical wiring problems, the FD50 ended up a casualty of certain fiscal changes in motion at Fiat during the early 1990s. Fiat-Allis also produced the 425bhp Fiat-Allis 31 from 1976, the 455bhp FD40 from 1982, and the 475bhp FD40B in 1989.

KOMATSU

Another intruder into the Caterpillar hegemony during the 1970s was

This Caterpillar D6H is working on one of the UK's most important construction projects, the Channel Tunnel, clearing sand at the beach terminal. Different blades can be fitted according to the nature of the material being moved.

Komatsu. This long-established Ishikawa-based firm offered its D50 series bulldozers in 1947, running with its own 60bhp 4D120 diesel engine. By 1970 its production tally of D50s had topped 50,000. In 1974, Komatsu came out with the D355A-3, and the D455A-1 emerged a year later. The 410bhp Komatsu D355A-3 scythed through the earth with a 4.6m- (15ft-) wide blade that could shift 16.8 cubic metres (22 cubic yards) of soil, and was fitted with the in-house SA6D155-4A six-pot diesel engine. Its power output equalled that of the Cat D9G, and it was seen as a deliberate attempt to cream off potential Caterpillar D9G/H buyers. In the same way, the hugely powerful 620bhp Komatsu D455A-1 was aimed fair and square at the Fiat-Allis 41-B. The big, angular Komatsu dozer with its ROPS cab was, for a short time, the largest tracked dozer in the world. It was fitted with a blade measuring 6.1m (20ft) wide that could handle 1.2 cubic metres (42 cubic feet) of overburden, and drew on a 12-cylinder Cummins VTA1710-C800 diesel engine as its power source. Meanwhile, Caterpillar retained its market advantage throughout the 1960s and 1970s with the D9 series, offering the 410bhp D9H model in 1974. Production carried on until 1981, during which time its all-up weight rose to 47.8 tons.

WHEN PUSH COMES TO SHOVE

If you want more power, you can nearly always double-up on power units, as demonstrated by the twin-engined models. Caterpillar proved that double the power could be available by twinning or tandem-harnessing its existing D9 dozers. The first such amalgam was the Quad-Track DD9G, introduced in 1964, which was basically two identical units linked one behind the other. The engines were the same as those fitted in the standard D90, and developed 770bhp. Two years on, the Quad-Track prefix was dropped and the designation was changed to the Cat DD9G. The 820bhp DD9H version replaced it in 1974, and this was in build until 1980. Originally thought up in 1963 by Caterpillar dealer R.A. 'Buster' Peterson, the Quad-Track was one of 10 units built for Caterpillar by Peterson Tractors of San Leandro, California, up to 1966. A further 51 Cat DD90s and seven Cat DD9H dozer sets were produced by Caterpillar itself after the patents were acquired from Buster Peterson, with a fair number of conversion kits for mating-up D9 tractor units also supplied by Peterson.

A variation on the tandem theme was the side-by-side configuration, known as the SxS D9G, which featured a pair of D9 tractor units and a blade twice the width of a standard one, at 7.3m (24ft) wide. Presenting a highly imposing façade, the SxS D9G was in production from 1969 to 1974, with 11 units made, and the SxS D9H was available from 1974 to 1977, during which time just 13 units were built. Predictably, the SxS D9H employed the same six-cylinder Cat D353 diesel engines as found in the standard D9H and the DD9H quad-track unit.

PARALLEL TWIN

The most ambitious parallel-twin dozer was put together in 1978 by the Russell & Sons Construction Company, and was known affectionately as the Double Dude. The two Caterpillar SxS D9Hs were equipped with a 14m- (46ft-) wide bulldozer blade specially made by the Balderson Company and described as a BRAP III Land Reclamation Plow. In order to be most effective, the blade was set at 45 degrees, and the tractor units were thus positioned obliquely rather than in parallel. They were attached to one another by means of a special rear tie-bar while the hydraulic lines were mounted in a sling between the tractors. The Double Dude used the standard Cat power units, and could shift some 10,703 cubic metres (14,000 cubic yards) of spoil in an hour.

Caterpillar's next move was to bring out the first of its high-sprocket drive dozers in 1977. This was the D10 which, with 700bhp at its disposal, made it among the most powerful machines around at the time. What was unusual about its design was the

A typical medium-sized crawler dozer was this Blackwood Hodge 5204 model, fitted with a basic all-purpose straight blade for general construction site preparation. More modern tractors have roll-over cab protection.

Soldiers of the US 5th Marine Division get some mechanical back-up as a tank-mounted blade creates a roadway on the island of Iwo Jima in September 1945. Bulldozers were a vital part of military operations at invasion bridgeheads.

elevated track, achieved by means of a raised sprocket above and to the fore of the back axle, triangulating the configuration of the track run. This had the effect of minimising shock loads, helping to reduce issues to do with roller frame alignment, and shedding the spoil that was potentially damaging to the drive sprocket teeth. The raised sprocket undercarriage was available in three configurations: the standard arrangement for general duties; the XL arrangement, which was slightly longer and incorporated an upper runner, making it the tool for finish-grading or fine-dozing work; and the shallow triangle XR arrangement when undertaking ripping, drawbar or skidding tasks. Although a couple of precedents existed in the 1920s for the elevated sprocket design, it was novel in 1977 and, at a stroke, the D10 changed the look of all Caterpillar dozers. It was powered by a 700bhp 12-cylinder Cat D346 diesel engine, which was turbocharged and aftercooled. Transmission was by way of a single-stage torque converter with an output torque divider. Seventy-five per cent of torque was dispatched through the converter, and the remainder via a direct driveshaft. The planetary powershift accessed three forward and three reverse speeds, utilising big, oil-cooled clutches. The blade at the dozer's business end measured 6.1m (20ft) wide by 2.1m (7ft) high, with a capacity of 26.7 cubic metres (35 cubic yards). While early models had a single exhaust stack, later versions had dual pipes.

The D11N, revealed in 1986, was an uprated version of the D10N. Darling of most major mining operations, the D11N used an 11 U bulldozer blade measuring 6.4m (21ft) by 2.3m (7ft 7in) high, with a capacity of 34.4 cubic metres (45 cubic yards). It was equipped with a new 770bhp eight-cylinder Cat 3508 diesel engine and its roller and track frames were extended to increase the tracks' contact patch. The D11N was itself eclipsed by the D11R model in 1996, which enjoyed a power boost up to 850bhp the following year. This version was fitted with electronic steering and transmission control, with a tiller or joy-stick operable if necessary with one hand to make life much easier for the driver. The 'accelerator' was simply a knob on the end of the tiller, and steering was by means of a hydraulically operated planetary differential. This ensured a smooth, consistent power delivery to both sets of tracks. The dozer blade could accommodate 20.3 cubic metres (26.7 cubic yards) of rubble, and was 5.25m (17ft 3in) wide by 2.1m (7ft) high. Caterpillar announced an evolution of the D11R at the 1996 MINExpo in Las Vegas, known as the CD or Carrydozer. This machine was powered by an 850bhp Cat 3508B diesel engine, and fitted with an extraordinary 43.5-cubic metre (57-cubic yard) blade that measured 6.7m (22ft) wide by 3.25m (10ft 8in) high. The concept was that the dozer could carry overburden within the curvature of the blade curvature, adding a slightly different dimension to the vehicle's capabilities, and meaning that it could propel a greater quantity of spoil ahead of itself. Launched in 1996, the Carrydozer was some 5 tons heavier than the normal D11R.

AMAZING ANGLES

By the late 1990s, Caterpillar was offering a range of 26 crawler dozers, ranging from its diminutive 7.8-ton D3C III model to the giant D11R that weighed in at 108.4 tons. Maximum speed of a smaller model like the 133bhp D6R was 11.5km/h (7.3 miles/h), but even the giants in the range were hardly any quicker. Tracked dozers could operate at amazing angles on inclined slopes in normal format, but low ground-pressure versions were also available, equipped with wider-track shoes, and longer and wider chassis frames to spread the vehicle's weight on boggy terrain. Accessories included multi-shank rippers, winches, and varying blade widths and patterns, such as the U-shape, straight tilt, and angle-dozer.

Probably the manufacturer which has presented most of a challenge to Caterpillar during the 1990s has been the Japanese firm Komatsu, contesting the supremacy of the Cat D11 models with its D475A machines. Komatsu came up with its D475A-1 dozer in 1987, relying originally on its own 740bhp eight-cylinder SA8V170 diesel unit for motive power, but gaining an

A shower awaits the driver of this Komatsu dozer moving silt from a basin at the Channel Tunnel site on the Kent coast. The 142bhp D65PX-12 has a reputation for good manouvrability in muddy conditions, and will be running broad tracks to spread the vehicle's weight on the sandy bottom.

extra four cylinders two years later with the D475A-2, running the 750bhp 12-cylinder SA12V140 diesel engine. The 34.4-cubic metre (45-cubic yard) capacity blade was 6.2m (20ft 4in) wide by 2.6m (8ft 7in) tall. A decade on, Komatsu was ready to reveal the latest in the series, its 860bhp D475A-3 SD model. This machine went to work with a 6.4m- (21ft-) wide blade that had a 45-cubic metre (58.9-cubic yard) capacity. But by comparison, the same company's massive D575A-2 SD – which came out in 1995 – was the largest crawler dozer on the market. The prototype of this machine was unveiled at Conexpo, Houston in 1981, where it was described as the D555A. However, before it could become a productionised reality, the advancing recession, coupled with low mineral values, called a temporary halt to development of this 1000bhp, 132-ton behemoth. Work began again in 1986, and by 1989 it was undergoing testing. The model was finally launched as the 1050bhp D575A-2 in 1991, the first unit off the production line being acquired by the American Asphalt and Grading Company, operating near Las Vegas.

THE TOUGH GET ROUGH

Not for nothing was the SD evolution that appeared in 1995 known as a Super Dozer. It was rated at a cool 1150bhp and brandished a colossal 7.3m- (24ft-) wide blade. It needed a specially designed front end to hold up the blade's hydraulic control cylinder, and an uprated and balanced chassis with rear-mounted counterweights. These dozers don't score too highly in the aesthetics department, but then, styling is hardly what counts here. Nowhere else does form follow function more faithfully. The D575A-2 Super Dozer (or SD) used the 12-cylinder Komatsu SA12V170 diesel engine to push along its 68.8-cubic metre (90-cubic yard) blade and 158-ton bulk.

Earlier D575A-2 models were known retrospectively as SR or Super-Ripper dozers, a reference to the powerful hydraulically operated appendage that protruded from its rear quarters and performed exactly that function on the earth's surface. The ripper shank, which was shaped like a claw, was particularly adept at demolishing dense rocks up to a depth of 0.9m (3ft), and was favoured in sites close to urban areas where blasting of overburden was banned.

In 1996 Komatsu unleashed its revamped D375A-3 model that had its origins in 1987 when it could muster some 500bhp. In revised format, the D375A-3's six-cylinder Komatsu

SA6D170E diesel unit pumped out 525bhp and furnished a blade capacity of 20.33 cubic metres (26.6 cubic yards), equalling Caterpillar's D10R. The dual-tilt dozer-blade facility also aided the machine's productivity, as the angle of the blade could be set for best effect on all kinds of overburden and gradients. Lower down the Komatsu range was the noless sophisticated D155A-3, powered by a 225bhp Komatsu S6D140 engine, and featuring the company's REU system, or Resilient Equalised Undercarriage. This meant in practice that the triple sets of bogies in the undercarriage moved independently of one another to help the tracks maintain contact with the ground surface over extremely bumpy terrain, and the bogies themselves were fitted with rubber shock absorbers to minimise vibrations. Like contemporary crawler dozers, Komatsu powertrains were equipped with electronic monitoring, and major componentry was modular in concept for ease of replacement and transportation.

THEY ALSO SERVED

By no means did Caterpillar, Fiat-Allis, and Komatsu have a monopoly on big, powerful dozers. The other players in the field were the heavy-plant equivalent of the one-night stand. However, only a single vehicle was made in most cases. These included the diesel-electric drive Doerr Tractor produced by the Charles Doerr Company of Canada, and delivered to the Mannix Company's Alberta Coal operation in 1966. It was built up on a 10m- (33ft-) long articulated chassis, using four Euclid TC-12 crawler assemblies, and was powered by two Cummins engines via a quartet of General Electric traction motors, with a combined rating of 984bhp. The Doerr Tractor tipped the weighbridge scales at 88 tons.

This 391bhp Komatsu D375A has its semi-U blade filled to capacity, and takes full advantage of its dual-tilt option, whereby the angle of its 4.5m- (15ft-) wide blade can be set to cope with all kinds of material and terrain inclination.

Another leviathan was the ACCO Dozer, unveiled in all its colossal splendour by its creator Umberto Acco at the Verona Construction Trade Show in north-eastern Italy in 1980. It was a tall machine by any standards, and at 183 tons, positively the world's heavyweight crawler dozer. It incorporated the Caterpillar high track format with drive coming from upper and lower sprockets on either side, derived from a pair of Caterpillar V8 diesel engines mounted side-by-side and developing a combined 1200bhp. Its blade was a relatively modest 7m (23ft) wide by 2.7m (9ft) tall, although it featured a ferocious-looking ripper spike out at the back. It was the fate of the ACCO Dozer to be specially commissioned for a substantial earth-moving project in Libya. However, it was stopped in its tracks by the international trade sanctions that held sway at the time against Libya and the order was accordingly cancelled. Astonishingly, no other buyers came forward, and the ACCO remained on the forecourt of its creator's premises as unique testimony to his creative talents.

When the International Harvester Company divested itself of its construction equipment division in 1982 to the Dresser Company, the TD line that dated back to the early 1930s IH tractors remained in production. It was manufactured in the company's Polish plant as the 460bhp TD-40B, which came out in 1985. The TD-40B that superseded it appeared in 1989 and was a conventional looking broad-cabbed dozer, powered by a 520bhp six-cylinder Cummins KTA-19C diesel engine operating with a standard blade width of 4.9m (16ft) and capacity of 18.8 cubic metres (24.6 cubic yards). By 1997, a revised TD-40C model was available, complete with a new ROPS cab, incorporating current safety features.

One manufacturer offering a range of equipment to the mining and construction industry was Liebherr, who made a line-up of five crawler dozers and four tracked loaders. The dozers included the PR712 and PR751 models, all of which used hydrostatic steering and transmission, producing an effect not unlike traction control. It enabled the machines to function without slipping and without the need to de-clutch. In the late 1990s, the largest Liebherr model was the PR751, a stocky machine powered by the 243bhp Cummins KT 19-C turbodiesel unit.

What about the life expectancy of a tracked dozer? The smaller sub-100bhp models should be good for 12,000 hours of use before a major rebuild is required, while the really big boys can soldier on for around 20,000 hours.

MOTOR SCRAPERS

When contractors set about a new road-building project, they're inevitably going to have to make changes to the landscape. This means building embankments in valleys and making cuttings through hillsides, and it's here, in major projects like these, that the scrapers and graders make their mark. Road-building is essentially the process of cutting material from one location and then depositing it in another. The contractor cuts through the hills or prominences, and deposits material in the valleys or hollows. This has the effect of easing the contours of the natural terrain. We've come a long way since the first European road-builders, the Romans, who famously built in straight lines, but who also picked the line of least resistance.

Twin-engined Terex scrapers working in tandem with Caterpillar pusher-tractors to level a site in southern England. Once the tractors have pushed the scrapers through the cut, they will return to do the same again for the waiting pair. The 'button' on the rear of the scraper is where the tractor pushes.

Now there are all sorts of restrictions that forbid us from doing what the Romans did, and we are sometimes obliged to cut through hills at base level, as it is the only route available. Apart from mining operations and big civil-engineering projects like hydroelectric dams, this is the main reason why we have large, earth-moving machinery.

The slopes that form the shape of embankments and cuttings – cut-and-fill areas, in muck-shifting parlance – are the 'batters', and are formed to very precise levels. In nearly all earth-moving operations, be it roads, civil engineering, or mining, the first operation is the removal of the topsoil or overburden. The depth of this soil can be anything from a few inches to a few feet or so, depending on the locality. The topsoil is saved in large heaps for future use. The most efficient machine for this operation is a scraper, as it can form the heaps, given competent drivers, without the aid of unproductive machinery. Again, in the trade, 'unproductive machinery' is anything that does not move material.

Once the topsoil has been put to one side, the serious business of making the road, constructing the dam, or excavating the mine, commences. All these operations involve moving subsoil. This, as distinct from topsoil, is anything below the organic, plant-supporting topsoil. Subsoil may be composed of granite, clay, gravel, chalk, alluvium or any of the other types of sub-strata. Each material has its own properties, which affect the earth-moving operation in some degree. Excavating rock and moving it to another location, providing it's hard rock, is an operation that runs independently of weather conditions. Clays and chalk, on the other hand, become slimy and impassable for rubber-tyred machines when it rains, so a job concerned with moving these sorts of materials is extremely weather-dependent.

Once the mineral sub-strata has been cut and put in place, the vista can be an extremely bleak one. The heaps of topsoil are then dug out, either by scraper or excavator, and taken by dump truck to the embankments, cuttings and tips for later use in areas deemed unsuitable

The first graders were drawn behind tractors, as demonstrated by this Royal Engineers' combination, grading a road through the Burmese jungle in 1944. The cambered road extended for hundreds of miles, surfaced with roofing felt to counteract the monsoon rains.

A Fiat-Allis 60B scraper working at Savoigne, Senegal in 1979, with bowl unit raised. It typifies the modern motor scraper, with single-axle tractor unit, swan-neck connector and bowl unit. Note the driver hidden behind the exhaust pipe.

for construction. The topsoil is placed over the bare sub-strata material that has been used in constructing the embankments and the slopes or batters of the cuttings. This has to be carried out to precise depths of topsoil, and the finished level must be exact.

SCRAPING THE BOWL

The earliest scrapers were, like their grader counterparts, simple, lightweight implements that were trailed behind a tractor. They consisted of two iron wheels, with a bowl mounted in between them, which was effectively a cylinder cut in half along its axis. This truncated cylinder was positioned so that its belly was at the rear of the machine and its open part at the front. The scraper bowl was set at a certain level and initially it cut high spots on the ground, filling the bowl with material, and then deposited the same stuff into the low-lying spots. When fresh air appeared underneath it, the material spilled out. It was quite basic. After a certain period of working, a level would be achieved, at which point the cutting edge could be lowered to take a continuous cut. The bowl filled up accordingly, and as long as it was in contact with the ground, it stayed full. When it was appropriate to dump the contents, the tractor operator pulled a cable, causing the bowl to rotate about its axle and dump the material.

The next development in scraper technology was to mount it on a couple of axles and tow it with a crawler tractor. The wheels were now shod with voluminous balloon tyres, and the rear ones were fixed directly to the rear of the chassis. In front of them was the box, which consisted of a compartment with two sides and a fixed bottom. At the front of the bottom part was a cutting edge. Fitted at the front between the two sides of the box was a door which could be raised vertically by means of steel cables. At the back of the two sides, the end could be moved forwards, towards the door, again by means of cables operated by winches on the tractor. In this way, the whole box could be raised and lowered about

the axis of the rear axle. Forwards of the door there was a swan-neck fitted to an axle. This axle carried the front wheels, and attached to it was a drawbar that coupled up to the tractor.

The scraper then evolved into a motor scraper. This was likely to have been a single-axle tyred tractor instead of an independent crawler tractor. The swan-neck now fitted to the centre of the front axle of the tyred tractor and above it. The box was still cable operated, but this time by winches on the tractor unit. Cables were eventually replaced by hydraulics, and an optional rear-mounted engine was frequently offered in addition to the front engine. The operation of all motor scrapers, hydraulic or cable, twin or single engine, is the same, with the exception of a category called elevating scrapers.

AT THE CONTROLS

To most people the operation of such machines has always been something of a mystery. Now, though, we get to have a driving lesson from Stubbsy (Gerald Stubbs is an expert machinery operator) who asks us to imagine we're in the driving seat of a Caterpillar 637,

Tea break: the green Terex scraper and Cat D9 tractor are temporarily rested. Although the scraper is self-propelled, it is common practice for it to be assisted in its cut and fill tasks by the bulldozer, which shoves it along from behind.

a mid-sized machine. It has two engines, the main one at the front to the right of the left-hand drive seat, and the other one, which is smaller, between the back wheels of the box. The single gearlever that operates both transmissions is next to the seat to our right, against the engine cowling. There are two throttle-pedals, set side by side, so that you're able to operate one or both of them with the right foot. To the left of them, though still operated by the right foot, is the brake pedal. The handbrake is a button, which is a pneumatic valve.

The driving seat is angled to the right, so that you are sitting slightly offset. This is because the operating levers of the machine are slightly behind you and to the right. When you're charging the machine, you'll be looking at the cutting edge of the box, which you can just see over the mudguard of the wheel

Scrapers at work during the creation of a dam at Ignacio, Colorado in 1967. A Cat tractor shod with compactor wheels pushes a Euclid scraper up the hill, while a water truck dampens the fresh lift. The Cat D8 blade in the foreground is removing wet excavation material.

behind you. Looking at the scraper from the left-hand side, your cockpit is just forward of the top-left quadrant of the front wheel.

The three operational levers are in a row along the axis of the machine, and are operated by your right hand. You have to be a contortionist to use the left one! The lever nearest the front of the machine is the box lever, and that raises the box up and down. Away from you is for down, towards you is for up, and the middle position that the lever is sprung to is 'hold'. The second lever is the apron (or front-door) lever. You pull it towards you for raising or opening the apron, push away from you for lowering or closing, and the middle position is, again, sprung for hold. In addition there's a further position – past 'down' – which the lever will stay in, which is 'float'.

The rear lever is the one that controls the ejector. If you open the apron fully, you can now see right into the interior of the box. Pull the ejector lever towards you, and you will see the back portion of the box start trundling towards you. This is the ejector. You let the lever go, and it will snap back to the middle position, which is 'hold'. Push the lever away from you and it will stay there, and the ejector will return to the back of the box. When it gets right back, the lever will snap back to the hold or middle position on its own. All these operations are hydraulic ones. You are operating hydraulic valves that direct oil under high pressure along pipes and flexible hoses to the large hydraulic rams. It is these rams that do the work.

HAVE A GO

You want to give it a try? It's far wider than anything you've driven before. It's also very long and you're a long way off the ground. First of all you put the transmission into fifth gear. Stubbsy advises limiting it to that until you start to get

Like antediluvian monsters emerging from a swamp, these laden Terex scrapers are off to the fill area to dump the contents of their boxes. Apart from the lead scraper, the three push-and-pull units following behind have bale-arm links to connect up for tandem work.

the hang of it. You have your right foot on the brake to prevent it moving off. When you put it in gear, the machine jumped a bit. Now you engage the cushion-hitch, which is operated by a little lever on the front of the gear console. You won't be prepared for what happens next, as the nitrogen cylinder comes in with a bang and the swan-neck linkage – a parallelogram affair – instantly moves about 0.3m (1ft), and the front of the bonnet rises up.

Then you lift the box off the ground, check that the ejector is back, and make sure that the apron is in float. If you let go the footbrake and press both throttles, the machine instantly surges forwards and changes up through the gears with some rapidity. Every time it changes gear, the swan-neck bounces up and down. We have to get to the cut, which involves doing an about-turn. You have to make a fairly tight turn here, so you must take your foot off the rear engine's throttle in case it pushes the tractor sideways when it's on full-lock. Now that the machine is straightened up again, put the rear engine back in.

As you approach the cut, you notice that the scrapers being pushed in the cut are working towards you. You keep out of their way and get to the back of the cut, where you must turn again to line yourself up. You see a pusher waiting for you, sitting at 45 degrees to the line of the cut. You need to pass just in front of it, so that he can pick you up. He's sitting at that point because that's where he wants you to dig. Don't go round behind him, or decide that you want to dig somewhere else, because you'll annoy him, and he will probably pick you up in third gear, which will put you out of your seat and bang your head on the fly-screen! Making your approach, bang out the cushion-hitch, open the apron so that you have a good view of the cutting-edge, select a windrow that is nearest to your line, and lower the box, so that as you're coming in, you just skim the ground and pick up the windrow.

You pass in front of the tractor and only when you have passed it do you start to put a bit of cut on the box. This will lead you in and also bring the machine down into second gear. Put the lever into second once the machine has put itself into that gear, and leave enough throttle on to just keep you moving, while closing the apron to allow material to enter the box through a gap between the bottom of the apron and the cutting edge. Now take your foot off the rear-engine's throttle. You feel a nudge from behind, so floor the front engine while giving the box a bit more depth. You may have given it a bit too much cut, as the pusher is struggling, so you raise the box a bit and you go through the cut easily.

You notice that the material is coming up in a mound in the box, so as soon as it starts to spill over the sides, whack the gear lever backwards, and steadily raise the box out of the ground so that you don't leave a step. You hit the rear-engine throttle so that both engines are working together again. Then you throw the apron lever into float and put in the cushion-hitch. All this time, you'll have been remembering to steer the scraper, naturally! You are now on your way to the fill area.

On arrival at the fill area, you throttle off a bit, and the scraper starts coming down the gears. Each time it changes, the box works the cushion hitch up and down. If it did not, the momentum of the weight of the box would be

working on the front tyres, making the scraper bounce. The cushion-hitch absorbs that and gives you a (relatively) less bouncy ride. You see that the embankment has risen from the surrounding area by some 9.1m (30ft) at its deepest point, which is halfway along the embankment. As there is no slasher to direct you to where to discharge, you give the surface of the embankment a quick glance to establish the way the fill is lying. As it looks pretty level, you must fill one of the edges. You are still getting the hang of the scraper, so you can run along the left-hand edge and tip there. This puts you right on the edge, as it's left-hand drive, and you'll be able to see exactly where you are in relation to the edge, instead of having to do it blind, as you would on the right-hand edge.

You lower the box to about a foot off the ground while you are coming in. When you get on the edge at the place where you wish to tip the load, you open the apron until material spills out of the front of the box. Don't open it any more than you need in order to achieve this, or you'll lose control of the flow. The material falls under the box, and the cutting edge roughly levels it. As soon as the back wheels come on to what you've tipped, you may have to lower the box a tad to achieve the same depth of fill. After the material that was resting against the back of the apron has fallen out, you start using the ejector to move the rest of the material forward so that, in turn, it falls out of the front.

When the box is empty, close the apron and put it into float, and put the ejector lever into its return position. When it gets back, it puts the lever into hold. If you want to be accurate with your levels, then it's best to put out the cushion-hitch when you're tipping. For bulk-filling, there's no need to do this.

CLOSE TO THE EDGE

When you're going along edges, you must always remember that if the side gives way and the scraper starts to slip over the edge, it's very important to overcome the natural reaction to steer back on to the fill, as this has the effect of putting the body of the scraper further over the edge. This gives the classic roll-over situation, resulting in an upside-down scraper at the bottom of the batter and, like or not, a squashed driver. The best thing to do is to turn the nose down the batter and drive straight down it, remembering not to use the back engine when turning. Use

R.G. LeTourneau was one of the chief innovators of earth-moving equipment. This triple scraper unit is the LT-360 of 1965, powered by no less than 8 turbocharged Detroit Diesel engines and 12 electric traction wheel motors. The driver's cab is behind the second pair of engines on the right.

the box to moderate your progress, and you can go down very steep slopes in perfect safety, so long as you go directly down the slope and don't take them at an angle.

The main reason that you always work the edges first is so that they are built up before the middle. The middle will always look after itself. If the edges are built first, then it's quite difficult to stay on the outside, as the scraper is inclined towards the centre of the fill. Roll-overs normally occur when there are inexperienced drivers in a field who are nervous of going to the outside, resulting in a hog-backed fill with low edges. In this situation, the scraper is naturally weighted towards the edge, and it's difficult to stop it slipping over.

ELEVATING SCRAPERS

There is another sort of scraper which can work on its own unaided by a pusher. This is the elevating scraper. They

MOTOR SCRAPERS

ELECTRIC DIGGER

Make: R.G. LeTourneau
Model: LT-360 Electric Digger
Manufactured: 1965
Engine: 2 Detroit Diesel 12V-71N & 6 Detroit Diesel 16V-71 turbodiesels
Power output: 4,536bhp combined
Bowl capacity: 360 tons
Length: 61m (200ft)

are used normally for topsoil stripping and other similar work, but not normally for bulk-shifting. Instead of having an apron, they have a chain the width of the machine, fitted with paddles that work against the load, piling it up in the box. When they tip, the floor slides back to allow the material out.

There's also a variant on the twin-engined scrapers, called push-and-pull scrapers. There must be more than one of them for the system to work, and there is usually an even number of them, as they work in pairs. A push-and-pull scraper has a pushing block fitted to the front of the machine as well as the back. Mounted on the rear pushing block, on the top of it and forward of the working surface, is a large, forward-facing hook. The front of the machine is fitted with a device called a bale-arm, which is a hinged metal section articulating either side of the machine on an arm that goes forward and forms a semi-circle, then comes back to make the other arm. When not in use, it is carried vertically, and when in use, horizontally.

Going into the cut, a pair of scrapers take the same line, one after the other. The first one grades-in and then starts to take his cut. The second one buttons-up to him and, using both his engines, pushes the first one through the cut, keeping his own box out of the ground. When the second one pushes the first one through the cut, he also lowers his bale-arm over the first one's hook. Then, when the first one is filled, he raises his box and closes the apron. The second one then picks up the cut where the first one had finished.

The first one, now using both engines, pulls the second one through the cut. At this stage there is no contact between the pushing blocks, the action being taken up via the bale-arm and hook. When the second scraper is loaded, he moves forward to button-up again, thereby taking the contact and tension off the bale-arm so that he can swing it up out of the way. At this point the scrapers act independently again, until they both get back to the cut, where the process repeats itself. There. Driving lesson over! Now go and sign up with a contractor as a scraper driver.

SCRAPING A LIVING

In reality, the route to a job behind the controls is relatively straightforward, but involves learning various other skills along the way. It used to be the case that to operate a bulldozer, grader or scraper required no qualifications. A new recruit was generally taken on by the contractor as a dump-truck driver, as this was the machine that required the least specialist skills. A driver who demonstrated ability, common sense, and an ability to read ground conditions – very often people with agricultural machinery experience, like young farm-workers – and who picked things up quickly would then be taught how to drive a motor scraper. From there they might qualify as a blade- or crawler-dozer driver and move on to a blade. Some dump-truck drivers might take the career route directly to hydraulic excavators and stop there.

The most skilful job in the earth-moving industry is driving a motor grader, more so than a scraper. A scraper driver would ride with a grader if he was idle, because of his machine being broken down, for example, and thus pick up the rudiments of grading the access road, known in the trade as the haul-road. He might be spotted doing this of his own volition by the general foreman and noted as a potential driver and then given the chance to move on to a grader. From there, he would progress to grading formation. That means making the finished earthworks level prior to construction operations starting. The grader driver would be expected to work to a finesse of plus 15mm (0.59in) and minus 30mm (1.18in). This would be checked by eye through sight rails, and there would be a fair amount of leeway, depending on the diligence of the inspector.

Having proved himself as an accurate operator, a grader driver would then go on to 'final trim' and trim up the sub-base of the road or building site to the desired level. The tolerance here is generally a stricter (plus) +10mm (0.39in) to (minus) -15mm (0.59in), making it pretty exacting work. The levels are checked with a string-line and ruler every 10m (32.8ft) along the length of the road, and at five or six points along its width, and then recorded. The driver's reputation is built on good 'dips', as string-lining is called.

As far as scraper drivers are concerned, the less skilled operators take care of the bulk work, which implies cutting-and-filling, where little accuracy is needed. A driver who displays a certain amount of finesse in this would be put to work doing ground formation in front of the grader, cutting to within 25–50mm (1–2in) of the grader's level.

This is actually quite difficult, as a scraper has only a vertical lift, so the cutting edge of the box – meaning the rear part of the scraper that fills with material – will be at an angle across the machine that is dictated by the attitude of the back wheels. Thus, if one wheel is standing on slightly high ground, the cutting edge may be leaning the wrong way to the desired position. For example, the driver may want to cut high ground on the left of the cutting edge but find that, when he gets to the ground that he wishes to shave off, the cutting edge is leaning the

A convoy of Terex scrapers prepares to go into the cut. Working in pairs, they take the same line, and as the first one grades in, the second buttons up behind and, with both engines running, pushes him through the cut. Crawler dozers perform a similar function and are known as 'pusher tractors'.

wrong way. It's here that an ability to see levels in the ground from the perspective of the driver's seat comes into play, and the driver can weave the machine around, taking advantage of the way the ground lies to achieve the required position. A scraper driver who becomes good at this is a potential grader driver.

Driving a blade or crawler dozer poses similar difficulties. A tractor that doesn't have a blade-tilt has to be driven in such a way to get the ground initially level, and the problem is the same as for the scraper as described above. However, once a level pad has been achieved, then the driver has a uniform base on which to work. This is regarded as high-class work, which includes trimming the sides of the embankments – batters – of a new road scheme, topsoiling, or even landscaping.

POWER SHIFT

In the last couple of decades, just about every application is operated by a power-shift system. In the case of a scraper, it's a two-pedal job. There are variations on the transmission theme, including epicyclic transmissions that work in combination with a torque converter. On the Caterpillar 631Cs and 631Ds for example, the transmission has eight gears. Reverse, first and second are torque converter gears, and in all the other gears the converter locks up and gives direct drive from the engine to the transmission. If the driver puts the gear into eighth, then the transmission will run automatically between second and eighth, assuming that conditions allow the machine to work its way into the latter gear. Assuming the scraper is pulling into the cut, then it will lose speed and, as it does so, the machine will drop down through the gears. The driver grades in with the cutting edge, then comes to a stop to wait for the tractor unit – which might be a Caterpillar D9, for instance – to button up to the rear of the machine and push it through the cut. A properly fitted-out pusher, as they are called, is fitted with a spring blade, which is the same width as the tractor.

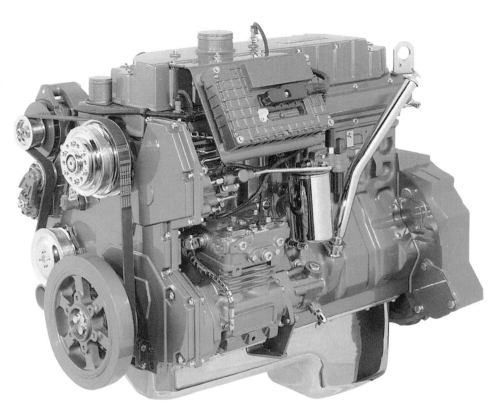

The Terex TS14D scraper is powered by two of these straight-six Detroit Diesel Series 40 engines, one in the tractor and one in the scraper, delivering a combined 234bhp. The low-emissions unit has a cast-iron block with wet liners and hardened crank, with service calls after 450 operational hours.

DETROIT DIESEL ENGINE

Make: *Detroit Diesel*
Model: *Series 40*
Number of cylinders: *six, in-line*
Manufactured: *1999*
Power output: *234bhp*
Block material: *cast iron*

This spring blade, hinged at the bottom and fitted with rubber bungs, is used only for pushing scrapers, as its use as a proper muck blade is limited.

When the scraper comes to rest, the transmission drops down into second gear, which is a torque converter gear and thus prevents the engine from stalling. The driver now has the choice of either selecting first gear – in which case it will stay in first, as the transmission won't step up past the gear selected by the lever – or, if the lever is still in the eighth gear position, the driver can step on the transmission lock pedal, preventing it from jumping into third gear.

When the scraper box is charged with material and raised out of the ground, the driver either shifts the lever from first to eighth gear, and it will automatically start on its way up the box, or if the vehicle has been in second gear with the transmission lock on, and the lever in eighth position, he can simply take his foot off the pedal. If conditions dictate that he doesn't want to go past fifth gear, then he just leaves the lever in fifth.

In very heavy-going work, there's another use for the transmission lock. When the machine is struggling to get through heavy clay, for example, it cannot make up its mind what gear to be in, and will continually jump up a gear, only to find the going too hard, at which point it will change down again, ad nauseam. In these conditions, as soon as the situation presents itself, the driver stamps on the lock when it's in the lower of the two gears.

ROLE OF THE PUSHER

The bulldozer driver has a key role to play in the activities of the scraper. The crawler dozer which we are concerned with here is known as a pusher tractor, because it's involved in pushing scrapers while they are loading material. The scraper driver must go to where the pusher is waiting and not make up his own mind as to where he thinks he should dig. The pusher also trims up the batters to the correct slope when he is not involved in pushing, so he is really running the cut.

When the pusher buttons up to the pushing block on the rear of the scraper, the scraper driver should not feel it, such is the finesse required. The pusher then lets the power in and, as soon as the scraper driver feels that, he then opens his own vehicle's throttles, and they go through the cut together until the scraper is loaded. At this point, the scraper box is raised out of the ground and moves away from the pusher. As soon as the pusher feels the pressure released, he puts the tractor in reverse to get back to pick up the next scraper which is waiting to load. This scraper would have graded in, picking up any errant windrows on the way.

A pusher may well have a sprung pushing block fitted to the rear of the

A motor scraper's box is operated by hydraulic winches, and the cut is made by the leading edge. A door at the front is raised to let earth in or out, and the back of the box moves forwards for unloading.

machine and be working in tandem with another pusher, similarly kitted out. This is clearly the most spectacular operation for an observor to watch, and it goes like this.

The first pusher picks up a scraper as described above, and the second then buttons up to the back of him. Then they both push the scraper through the cut. As soon as the pressure eases a bit, because the box of the scraper is coming out of the ground, the rear pusher reverses out of the way as fast as he can go. This leaves the first pusher to keep in contact with the scraper and help it move off, as it disengages from the ground by cogging up. By now, what was the rear machine is now buttoned up to another scraper. He is then quickly buttoned up to by the first pusher, who has arrived back at the beginning. And so they go on, leapfrogging from one scraper to the next.

The advantage of this 'tandem-pushing' is that the scrapers are loaded very fast and also that, if the cut suddenly fills up with scrapers waiting to be loaded, then the pushers can work singly until the bottle neck of scrapers is cleared.

When twin-engined scrapers are being used, it is also worth watching. First you get a cloud of black smoke from the first pusher followed by two plumes of the same from the scraper's engine, and then a fourth from the second pusher, all in quick succession. The smoke is produced as a result of the turbocharger lag.

THE GODFATHER OF SCRAPERS

Scrapers and graders are synonymous with the vision of one man: Robert Gilmour LeTourneau. He was a larger-than-life character who thrived on designing and building progressively bigger and better earth-moving machinery. By the time of his death in 1969, he held 400 patents for various types of plant.

A self-taught mechanic, R.G. LeTourneau was born in Vermont in 1888, dropping out of school at 14 and ending up in Stockton, California in 1909, where he later married, and went on to part-own a garage. By 1920 the business had failed, and LeTourneau set to repairing Holt crawler tractors and scrapers. Having bought his own equipment, he began working in the muck-shifting business and, on the premise that he could make a better machine than the Holt system, he constructed his own drag-scraper out of second-hand components. The second such vehicle was a 4.6-cubic metre (6-cubic yard) machine known as the Gondola, which was constructed in a novel way using brazing techniques as opposed to rivets.

The third machine off the LeTourneau line was the 9.1-cubic metre (12-cubic yard) Mountain Mover, which featured a two-bucket bowl layout for faster filling. His first diesel-electric machine was a self-propelled scraper with traction motors mounted in its steel wheels, introduced in 1923. As his earth-moving business developed, LeTourneau was involved in

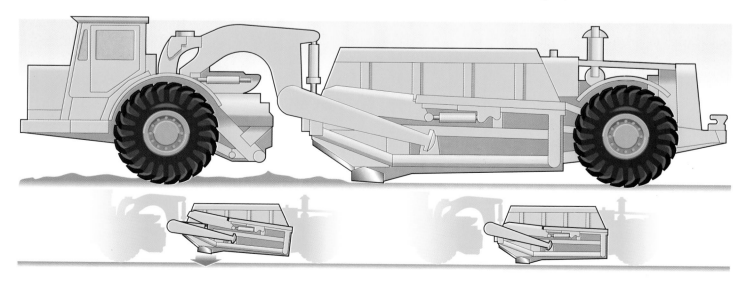

The bowl of the Caterpillar 611 is a relatively small unit, but serves to illustrate the business end of a scraper. As the machine prepares to move into the cut, the driver lowers the leading edge of the bowl into the earth, and raises it when it's full up.

some major US construction sites, such as the Boulder Highway project in Nevada, and by 1933 he was ready to start building his own vehicles from scratch. First up was the 6.8-cubic metre (9-cubic yard) Model A Carryall scraper. A factory was set up at Peoria, Illinois in 1935, and a further five plants were built in the US and Australia during the following decade. The first model to make a real impact was the self-propelled, rubber-tyred scraper-tractor, known as the Tournapull. This was a single-axle unit, powered by a Caterpillar diesel engine and hauling the Z25 Carryall scraper box.

Largest of the early mechanical, self-propelled wheel-scrapers built by LeTourneau was the Model A6 Tournapull, linked to an OU Carryall scraper. It consisted of a bull-nosed two-wheel tractor unit, with the engine compartment ahead of the axle line, Carryall hopper and blade, with dumping mechanism. The A6 Tournapull and OU Carryall were introduced in April 1942. The clutch-steered A6 was powered by two supercharged Cummins HBISD-600 diesel engines producing some 350bhp in total, mounted side-by-side, and mated up to a LeTourneau-designed power-shift automatic transmission. The double-bucket OU scraper could accommodate up to 46 cubic metres (60 cubic yards) heaped, and this was a considerable achievement at the time.

LeTourneau was at the forefront of scraper and grader technology right up until the late 1960s, and his company was responsible for supplying three-quarters of all the specialised earth-moving equipment used by the Allied forces during the Second World War. Among these were some 2000 Tournapulls, which were two-wheeled tractors that hauled scrapers, plus an estimated 15,000 bulldozers, 2000 cranes, and sundry buckets, shovels, and power-control units.

Although LeTourneau sold the rights to much of his earth-moving equipment business to the Westinghouse Airbrake Company (WABCO) in 1953, the revised company, known as LeTourneau-Westinghouse, continued to make the Tournapull and Carryall scraper units. When the moratorium restricting his new product development expired in 1958, LeTourneau astonished the muck-shifting community with a barrage of new diesel-electric powered machines using wheel-motor propulsion. Starring at the American Mining Congress event in San Francisco in 1958, the 38.2-cubic metre (50-cubic yard) Model A-4 Goliath, capable of hauling a 70-ton payload, seemed huge at 18.2m (60ft) long. This vehicle was powered by a 550bhp Cummins diesel unit, mounted way out front of the single axle. But LeTourneau had bigger fish to fry.

During the next couple of years he developed the Electric Digger range, which included the Pacemaker L-130, L-140 and L-60. Possibly the most successful of the Electric Digger line-up was the Series L-90, launched in May 1964. It was derived from the triple-hopper Series L-70 scraper, and powered by three Detroit Diesel 12V-71N engines, developing 1323bhp and capable of handling up to 53.5 cubic metres (70 cubic yards) heaped. It was available with electric-traction wheel combinations of 7, 8, or 12 wheels, and the overall length of the 12-wheeler version was getting on for 33m (108ft).

By 1965, the range had reached the LT-120, which had a scraper box that was capable of accommodating 55 cubic metres (72 cubic yards) of material. It was powered by four Detroit Diesel engines developing a whopping 2100bhp. Soon afterwards, a second scraper unit was added to the back of it, along with another three engines, and the designation changed to LT-240. Brake-horsepower rose accordingly to 2700bhp, while its carrying capacity was rated at 110 cubic metres (144 cubic yards). When a further scraper unit was added in 1965, the model name was changed to the LT-360. No fewer than eight Detroit Diesel engines were fitted, with two 12V-71Ns in the front tractor unit, and three 16V-71Ns mounted on the scrapers, totalling a staggering 4536bhp. An operator's cab was positioned – rather like a lookout tower – high up to the

right at the rear of the leading scraper bowl. Operational capacity was 165 cubic metres (216 cubic yards), which translates into 360 tons. The LT-360 was pressed into service on part of the Interstate 2 project in east Texas, along with other LeTourneau equipment. Back at base, the vehicle was subsequently re-evaluated, and the main drive wheels were reduced to eight in number in a bid to lessen its ponderous turning radius. The LT-360 was also fitted with eight Detroit Diesel 16V-71 engines, pushing out 4872bhp via its eight traction-wheel motors, and the model was relaunched in 1966. Impressive it may have been, but it wasn't as simple as that, as contractors decided it was just too large for convenient operation in this 61m- (200ft-) long configuration.

The multi-unit Electric-Diggers' final shot was the LT-300, which first appeared in 1966. The leading scraper unit accommodated 76.4 cubic metres (100 cubic yards), and the second hopper carried 55 cubic metres (72 cubic yards). The seven Detroit Diesel 16V-71 engines summoned up a stonking 4263bhp, driving six electric traction wheels. The cacophony thus produced would have sounded fantastic. A certain

This Cat 627 scraper working on the Kielder Dam site in the Scottish borders in 1976 shows the single-axle tractor unit and parallelogram swan-neck articulated linkage, plus the cables that operate the vehicle's hydraulics.

amount of cobbling up was done, because the rear hopper of the LT-300 was actually the third unit of the redundant LT-360, now with another engine bolted on. What remained of the LT-360 was designated the LT-240, now trimmed to six 16V-71 engines belting out 3654bhp. The Electric-Digger series evolved from the

MOTOR SCRAPERS

Longview crucible and, to an extent, represented a succession of bits and pieces of the same vehicle. The LT-120, LT-240, LT-360 and LT-300 all saw active service as prototypes, but their ultimate fate was to be broken up.

As R.G. LeTourneau experimented with the huge Electric Digger scrapers during the 1950s and 1960s, his original Tournapull and Carryall scraper lines were marketed by LeTourneau-Westinghouse. His only role in this activity was as a consultant, however. Some of the larger scrapers produced by LeTourneau-Westinghouse were factory-built tandem Model B Tournapulls linked to Model B Fullpak scraper units. The big tandem B Speedpull, or B-Pull, as it was known, came out in 1960, and could cart 48 cubic metres (64 cubic yards), provided it was heaped up. This machine was the larger sibling of the Model C Speedpull. Three years later, the designation of the scraper model line was altered to the WABCO Model B-70 Tournapull.

TANDEMS AND TRIPLES

At this time, the 44.3-cubic metre (58-cubic yard) capacity Model B Speedpull tandem and Fullpak triples were the biggest LeTourneau-Westinghouse conventional tractor-pulled scraper units. They were conventional in the sense that the tractor unit had two axles, and the entire rig, including both hoppers, was 23m (76ft) long. The power unit was a 550bhp Cummins VT-12 turbodiesel and if a further Fullpak scraper was added to form a triple combination, the carrying capacity was lifted to a heaped 66.5 cubic metres (87 cubic yards), or a payload of 102 tons. In 1961 it could also be specified as a single scraper bowl mated to the Model BM9 tractor, and rated at a maximum 34.4 cubic metres (45 cubic yards) or 52-ton payload. As an example of one contemporary commission, in 1960 LeTourneau-Westinghouse built a special triple set consisting of a Model B Speedpull tractor with Triple B Fullpak scrapers for the Healdsburg, California-based contractor Guy F. Atkinson. Along with many other major contractors and their scraper teams, Atkinson was involved on the San Luis Canal aqueduct project in California. The capacity for the entire 31.7m- (104ft-) long Speedpull-Fullpak ensemble was 66.5 cubic metres (87 cubic yards) heaped.

Hydraulics were still a little way off, and these machines all used electric cable methodology to activate the scraper bowl controls. In fact, there was little variation from the LeTourneau system which dated back to 1947.

However, this didn't deter customers, and WABCO enjoyed a measure of success with a line of self-loading elevating scrapers. Among these was the Model B-70, which featured a 23.7-cubic metre (31-cubic yard) capacity Hancock 333 elevating scraper, announced in 1963. Also in 1963, WABCO carried out experiments with a Model 800 Speedpull tractor hauling a tandem of Hancock 444 elevating scrapers, but the tractor was defeated by their combined capacity of 69 cubic metres (90 cubic yards), and it remained a prototype. Four years later, the 26-cubic metre (34-cubic yard) Model BT-333F came out, and this developed 885bhp from its twin-engine format. It was a long-running success,

A pair of Caterpillar 657E scrapers fill their bowls in a cut while a D10 blade operates on a higher terrace. The 657 series goes back to 1962, and is still regarded as the class leader. Its two Cat turbodiesels develop a combined power output of 950bhp, with bowl capacity rated at 33.6 cubic metres (44 cubic yards) heaped.

and not until 10 years after its launch in 1977 was the sizeable WABCO 333FT – as it had become – replaced by the even grander twin-engined Model 353IT. Summoning up an impressive 966bhp, this 27-cubic metre (36-cubic yard) self-loading elevating scraper was the largest such vehicle ever made. The last WABCO scraper left the factory in 1981.

ARRIVAL OF THE RIVALS

Before it became a division of General Motors, Euclid Road Machinery brought out the industry's first twin-engine scraper unit in 1949. It was pulled by an FDT two-axle tractor, and went into series production in 1950. The revolutionary twin-engined Euclid TS-18 scraper was unveiled in 1954, featuring a single-axle tractor unit with overhung engine in front. There was another engine mounted in the rear scraper unit, enabling it to undertake duties that single-engined scrapers couldn't manage.

WABCO's competitors in the scraper market of the 1960s included Allis-Chalmers' twin-engined tandem scraper model 562. It came out in 1962 and was capable of accommodating 67 cubic metres (88 cubic yards) of heaped material, while International Harvester's twin-bowl IH-295 PAY Scraper carried a slightly lower 48-cubic metre (64-cubic yard) rating. Euclid also made a couple of its biggest scrapers to special order for the Western Contracting Corporation. First up in 1963 was the TSS-40, a two-axle, tractor-pulled scraper that ran with two engines delivering 730bhp, and a bowl capable of holding 40 cubic metres (52 cubic yards). A year later, Western Contracting commissioned the Tandem TSS-40, which used three Detroit Diesel 16V-71N engines, rated at a sturdy 1690bhp. Together, the tandem scrapers' capacity totalled 79 cubic metres (104 cubic yards) maximum, or a 125-ton payload, and the Western commission extended to a further four units of the TTSS-40.

It was always going to be only a matter of time before the Caterpillar Tractor Company entered the self-propelled scraper market. The two-axle tractor DW20 and single-axle tractor DW21 existed in prototype form in 1948, both as single-engine units, and they were introduced rather tentatively in 1950 as further development was really required. Caterpillar took another 12 years to get around to launching its first two twin-engined scrapers. It came out with the Cat 657 and Cat 666 models in 1962 and, fittingly, both were major players in the big league. The single-axle tractor Cat 657 could originally muster 785bhp, rising through subsequent evolutions of the model to 950bhp in the late 1990s. While power increased, capacity remained at a consistent maximum of 34 cubic metres (44 cubic yards), and the fact that it was still in production some 35 years after its introduction suggests that Caterpillar got its sums right in the first place. Indeed, such was its build quality that many early 657 models were still in operation in the 1990s. The 1999 Cat 657E model was considered the industry class leader, and the tractor unit was powered by a Cat

MOTOR SCRAPERS

3412E diesel engine, with a Cat 3408E engine located in the scraper. An optional Coal Scraper bowl lifted carrying capacity up to 55 cubic metres (72 cubic yards) of this lighter commodity for use in mining applications.

The Cat 666 series was Caterpillar's biggest self-propelled scraper, measuring 17.2m (56ft 8in) long, and progressed by means of a two-axle tractor, carting 41.3 cubic metres (54 cubic yards) of heaped material in its scraper bowl. The 666's tractor unit was fitted with an eight-cylinder Cat D346 engine, and a six-cylinder Cat D343A was housed at the back of the scraper unit. It gained a reputation for rapidity, but in spite of its size, the 666 could at first only come up with the same power ratings as its smaller-capacity sibling, although the subsequent 666B model that was released in 1969 was endowed with 950bhp. This version remained in production until 1978.

Aesthetics are low on the list of priorities for scrapers. This Terex TS14C is equipped with a rudimentary roll-over protection canopy system to protect the driver if the scraper goes over the edge of a batter – as happens.

FAR CANALS

The Californian San Luis Dam Canal project aroused much interest among contractors in 1964 and, inspired by the prospect of a commercial gain, Caterpillar and its supplier Buster Peterson of the San Leandro-based Peterson Tractor Company rushed through a special tandem-rig scraper based on a Cat 657. The tractor and scraper bowls were linked and controlled by a solitary operator, using special controls that integrated the steering and accelerator systems for the four engines that were fitted. The 67-cubic metre (88-cubic yard) outfit performed well in trials at the canal's Third Reach excavations, and Peterson went for an even more ambitious triple-scraper bowl set-up, based once more on the Cat 657. The driver was perched in a cabin that looked like a sentry box, mounted above the rear of the first scraper unit, from where he could control the entire operation of the machine. The three tractor units ran with eight-cylinder Cat D346 diesel engines, and each scraper unit had a six-cylinder Cat D343A. Power output was 2580bhp, and maximum payload capacity of the three high-sided scraper hoppers was 115 cubic metres (150 cubic yards). If all you desire is magnificent overkill, there's no substitute for big stuff like the 657 Peterson Triple. Certainly contractor Peter Kiewit thought so, as he took delivery of the machine in 1965 and put it to work on the San Luis Canal at Coalinga. After a couple of years, this machine having shifted some 16.8 million cubic metres (22 million cubic yards) of dirt, Kiewit bought another Cat 657 Triple off Peterson. This second combination was subsequently broken down into a pair of tandem scrapers, retaining the single operator facility.

TEREX

Make: *Terex Equipment*
Model: *TS14D-II*
Manufactured: *1997*
Engines: *two Detroit Diesel Series 40 turbodiesels*
Power output: *234bhp combined*
Transmission: *Funk DF 158 Powershift, 7f/1r*
Heaped bowl capacity: *15.3 cubic metres (20 cubic yards)*

MOTOR GRADERS

You're travelling slowly along a busy road, cursing the stop-start traffic jams, and you notice that off to one side, they're building a new superhighway that some day will ease your pain. Chances are, among the heavy plant active on the new motorway scheme will be graders. Gawkier than scrapers, they are related by virtue of their enormity and ability to remove topsoil. There they part company, as the grader's job is to fine-tune the path the scraper has determined. Along with compactors, graders create a solid base on which to build roads. The opinion of operators of this machinery is that these are the tools that perform the fine scalpel work of the muck-shifting industry. The French have a different name for motor graders: they call them *niveleuse*, which means levellers.

If the bulldozer is a chisel and the scraper is a plane, the motor grader is the construction industry's scalpel, as it performs the fine detail work of levelling roadway and building site foundations. Here, a Caterpillar 126 is at work near Ramsgate on the UK Channel Tunnel project.

MOTOR GRADERS

Strictly speaking, a grader as we know it today goes by the name 'motor grader'. As a measure of the finesse sometimes required, the John Deere 772 motor grader has a laser unit mounted on the blade to chamfer the dirt with spot-on accuracy. The earliest graders were towed implements, with no hydraulic assistance. They consisted of an A-frame attached at the apex to a towing vehicle, with wheels at the bottom of the triangle. Mounted underneath the frame was a blade which could be orientated at an angle to the ground and raised or lowered independently at each of its extremities. These old machines had a driver on board, whose time was spent frantically turning the handles that operated screwed rods to raise or lower the parts of the blade as required. The angle of the blade was altered by pushing it round to the required position and then locking it in place with a dog or latch.

The modern motor grader is the development of that early machine. They're mostly six-wheeled, with a double-axle four-wheel arrangement at the back, underneath the engine. Forward of that is the operator's cab, and then there's a single-member box-section chassis that extends way forward and then down to support the single front axle. In the space between the cab and the front axle, and beneath the box-section chassis member, is the blade and the associated functional attachments that operate it. The blade itself is mounted on to what's called the circle drive. This is a circular ring within a ring, driven by an hydraulic motor, which enables the blade to be orientated at almost whatever angle the driver requires, such as a horizontal plane, except for approaching fore and aft.

The ring itself is fitted to the back end of an A-frame mounted at its apex, just behind the front axle. The mounting is a ball-and-socket joint, so the A-frame can adopt any attitude within the design limits of the machine. The perameters of these design limits are reinforced by physical stops which, once contacted, force the hydraulic system to open up relief valves. It's worth noting that all hydraulic machines have relief valves in every hydraulic line, which are designed to pop open if too much is demanded of the machinery. In this case, the oil is returned to the hydraulic tank so that the pump isn't asked to do more than it was designed to accomplish.

GRADER CONTROLS

The A-frame can, theoretically, do its own thing, but of course, the driver needs to be able to place it where it's wanted. To do this, there are two blade-lift levers, one for each side of the A-frame. They operate rams that are attached to the A-frame on either side of it, mounted vertically on gimbals. The rams are double acting, so one ram can lift or force down. That function enables the driver to obtain an attitude on the blade relative to the grader, which is essential because, of course, the

CATERPILLAR 24H

Make: *Caterpillar*
Model: *24H*
Manufactured: *1999*
Engine: *12-cylinder Cat 3412E*
Power Output: *500bhp*
Blade width: *7.3m (24ft)*

The largest motor grader in production at the new Millennium was Caterpillar's 24H model. Although its principal function was the grading of mining haul roads, its 7.3-metre (24-feet) wide blade, or mouldboard, could level two motorway lanes.

MOTOR GRADERS

Komatsu's 1990s models were the latest in a line stretching back to the horse-drawn graders of the nineteenth century. Pioneers of the motor grader were Russell and Galion, and by the 1930s, Caterpillar and LeTourneau had entered the market.

blade is fixed to the circle drive which is mounted on the A-frame.

Another ram is mounted horizontally, attached at one end to the chassis, and at the other to the A-frame. This ram's job is to swing the A-frame from one side to the other, and the implication is that the blade follows suit. The ram works at right angles to the longitudinal axis of the machine. Two rams are fitted to the blade itself. One of them will slide the blade from side to side, subject to its stops – which means the length of the ram action – while the other tilts the whole blade forwards or backwards. The former enables the driver to negotiate obstacles without having to avoid them by using the steering. The latter is a tad more complicated, and at this stage it's prudent to describe the usage of the motor grader.

It normally works with the blade set at an angle, picking up material and rolling it in front of the blade. The material itself travels to the trailing edge of the blade and escapes at the end, forming a windrow on the ground. If the blade is set back – and it's the top of the blade that is actioned, since it pivots at the bottom – the driver is in the optimum position for getting the material to flow along and off the blade. If the blade is set forward, then the machine crowds the material, so it cannot roll or flow on the blade, and the grader is able to carry it forwards, rather than have it disappear off the trailing edge.

As the grader is normally operating with the blade in an upright position so that the material flows, the cutting edge wears most at that angle. If the driver comes across particularly hard material that the blade can't penetrate, he will tilt the blade forward slightly to present it to the ground like a sharp knife, which will penetrate the solid stuff more effectively.

The side-shift ram, which pushes the A-frame from side to side – as opposed to the blade-shift, which fires the blade from side to side – does have its limitations. That limitation implies the stroke of the ram. On Caterpillar graders there is an additional facility for getting the A-frame slung out to the side of the machine, so that on the level, the grader is able to cut a slope to one side of it. This is achieved by a hydraulic pin that can be inserted into a number of different positions or holes, and acts on the mounting that the side-shift ram is fitted to.

GOING ROUND IN CIRCLES

With such a long vehicle, steering in circles is a problem. With as many as 10 hydraulic levers to operate, sometimes pulling five of them simultaneously, it is impossible for the driver to get to the steering wheel at the same time. This problem was addressed by Caterpillar in two ways. In the first instance, the front wheels lean, a function controlled by a lever. If the driver leans the wheels over to the left, then the machine will go to the left, and vice versa. Thus, without taking his hands off the bank of levers either side of the steering wheel, the driver could steer the grader satisfactorily. Any directional adjustment could be made when he had a hand free to nudge the wheel.

With the old rigid-chassis graders, a driver was obliged to make a lot of shunts to get the machine facing the opposite way, often with restricted space in which to do so. Caterpillar solved this situation by articulating the

MOTOR GRADERS

After a dump truck has deposited lines of top dressing, the Caterpillar grader gets to work spreading and levelling it across the whole site. This machine is also fitted with a multi-shank ripper, used to break up the ground prior to grading it.

grader between the cab and the engine and rear bogie unit. So when the driver wanted the machine to perform an about turn, he could put it on to full wheel-lean in the appropriate direction, give it full-lock on the steering, and hit the articulation on to the bump-stops. This was a great improvement on the old machines; they had wheel-lean and steering only.

As an optional extra, some graders are fitted with a scarifier, a set of tines mounted on an A-frame in an arrow formation, and fitted between and below the main A-frame that carries the blade and the front axle. The scarifier is used to break up the ground prior to using the blade on it. Also available is a ripper, fitted to the rear end of the machine, which serves the same purpose as the scarifier, but is a much heavier piece of kit, and is capable of penetrating material that a scarifier wouldn't have much effect on. The ripper normally has three or five shanks, with the three-shank ripper being a more serious item. Sometimes you can see a small bulldozer blade mounted in front of the front axle of the machine. This is for specialised grading operations but, according to our man Stubbsy, it is useful for lots of applications.

GETTING GOOD GRADERS

The interior of the grader cab features a comfortable driving seat. From here the operator commands the functions and console, which is hinged at the bottom, at the floor. This allows him to push the whole console forward and out of the way, for access in and out of the cab. A lever at the side swings the whole thing towards him so it's virtually in his lap. At the centre is a small steering wheel. Mounted on two shafts

An Aveling-Barford motor grader working with a road roller from the same Grantham-based company, levelling the foundations of a new road scheme. The trail of stones left by the trailing edge of the blade are known as windrows.

MOTOR GRADERS

New Holland's 170bhp RG 170 grader has an operating weight of 16,200kg (35,710lb) and a blade width of 4.3m (14ft). In 1999, New Holland N.V., which owns O&K, amalgamated with the Case Corporation, while parent Fiat retained a 71 per cent controlling interest.

Using the bank levers in his cab, a Komatsu grader driver demonstrates how the blade can be hoisted at an angle to level the batters or banking at the sides of road construction sites.

at either side of the wheel are small hydraulic levers – four on one side, five on the other – to operate rods that go down to the floor into spool valves. The driver's hands rest on the levers, and he can easily operate more than one at a time. There is an individual lever for hoisting the blade from left to right, wheel-lean, side-shift, blade-shift, blade lean, articulation, ripper or scarifier, and circle-drive. The bank of levers nearest the driver has three either side, and the other levers are set in another bank slightly forward, although mounted on the same shaft.

Just to the right of the seat there is the gear lever. This operates an epicyclic gearbox, but with no torque converter. The grader has direct drive in all gears, unlike a scraper, which has a converter between the engine and gearbox. There are six gears, forward and reverse. Like bulldozer tractors, there is one slot that runs from neutral to six (except that tractors only have three) and in neutral, the driver fires the lever across into the other slot to select the opposite direction.

On the floor, to the left of the main console, is the 'operating pedal' or clutch. It isn't a clutch pedal as such, as there is no clutch. It operates a spool-valve that hydraulically breaks the union between the engine and gearbox. On the right of the console is the foot brake that operates a pneumatic air system. To the right of that is the foot throttle or decelerator. This is a two-part item. There are two sections that stick up through the floor which the driver can operate with his right foot, and they are joined together under the floor, pivoting mid-way between the two pieces. The front part is operated by the toe, and the back part by the heel. If the hand throttle is

Among Bell Equipment's construction vehicle range is the 670C six-wheeler motor grader, seen here coming down a haul road. Note the blade's circle drive and A-frame located under the box-section chassis arm, and the hydraulic rams that orientate the blade at the required angle on either side.

set to full revs – as it would be if the driver was grading a haul-road – then the back part of the foot throttle is used, acting as a decelerator. If the hand throttle is closed down, then the driver uses the front part of the foot throttle, just as you would in a car. This would be the case if he was tackling any sort of tricky work done at low groundspeed. Grading-up haul-roads is done at the fastest possible speed, sometimes approaching 48.2km/h (30 miles/h), while cutting formation or doing final trim is carried out at a snail's pace. They constitute very different jobs.

Sitting in the grader seat, the driver is actually unable to see the business end of the machine, which is the cutting edge of the blade. He will be able to see what comes off the trailing edge of the blade, and what mark he is making on the ground with the leading edge, but that is all. The rest comes with experience, learned the hard way, and by a natural flair for the particular jobs a grader can do. Given those two abilities, the driver will know instinctively when to put an acute angle on the blade or when to operate it at right angles to the machine, when to crowd the blade or when to have it right back. He will feel through the seat of his pants what is happening, as well as what the attitude of the machine is relative to the blade and what angle the entire machine is relative to the ground. He'll know his machine so well that he'll be able to perceive that the ground 15m (50ft) in front of him is 25mm (1in) high. Even driving a

MOTOR GRADERS

AVELING-BARFORD

Make: *Aveling-Barford*
Model: *Aveling-Austin 99H*
Manufactured: *1960*
Engine: *9.8-litre, 6-cylinder Leyland AU600/14 diesel*
Power Output: *115bhp*
Blade width: *4m (13ft)*

dumper truck requires some skill, principally that of being able to read the ground conditions from the cab. Some good dump-truck drivers never, ever, get planted, while others, doing exactly the same job, get bogged six times a day.

HISTORY OF THE MOTOR GRADER

If the explanation of how a grader works seems close to awe-inspiring, bear in mind that they haven't always been so sophisticated. Back in the 1800s, a grader was nothing so much as a plank of wood fitted underneath a horse-drawn cart that smoothed away the wrinkles in prospective road surfaces. The first attempt to produce a machine that could operate on the sloping sides of a highway embankment was J.D. Adams' Little Wonder of 1885. Fitted with adjustable leaning wheels, the grader could be used more effectively on sloping sides of roadways. Adams got it right first time, as this leaning-wheel feature is present in all modern grading machines.

The prototype for the industry's first self-propelled grader was built by the Russell Grader Manufacturing Company in 1919. It consisted of a single axle, two-wheel Allis-Chalmers tractor unit up front, linked to a grader at the rear. However, the inaugural production motor grader was Caterpillar's Auto Patrol model, which came out in 1931. The Auto Patrol No. 9 was

An Aveling-Barford grader doubles up as a snow clearer, using its front blade as a one-way plough. As well as the normal mould-blade, it was regular practice to fit graders with big V-shaped ploughs on the front for deep snow clearance, and occasional derrick-mounted snow wings beside the cab.

built as a rear-drive machine at first, but from 1934, its No. 10 and 11 models were equipped with tandem-drive rear axles, which has become an element of the modern grader.

The grader established its credentials during the extensive highway-building operations in the USA and airfield construction during the Second World War, growing larger in the process. By 1963, Caterpillar's grader line had reached No. 16, and this machine was fitted with a 4.2m- (14ft-) wide mould-board blade and powered by a 225bhp Cat Diesel engine. The mouldboard blade was subsequently replaced with a broader 4.8m (16ft) version. Like many of Caterpillar's products, the formula was pretty much accurate from the

Operating a motor grader is a highly skilled business. From the cab, the driver of this Caterpillar model would be unable to see the blade in action, although he would see the windrow deposit coming off the ends, and he'll know from experience how much angle to put on the blade.

MOTOR GRADERS

outset, and not until 1973 did it come out with the 250bhp Cat 16G. A further 20 years would elapse before the 275bhp Cat 16H was unveiled in 1994.

As R.G. LeTourneau had experimented with used components to construct his early scrapers, we find South Dakota Euclid dealer Ken Harris doing much the same in the early 1960s. He called them Harris Blades, and they were made out of second-hand Euclid scraper- and TC-12 dozer parts. Towards the end of the decade, the Harris cause was taken up by RayGo Incorporated of Minneapolis, Minnesota. RayGo refined the Harris Blade to an extent and launched it in 1969 as the Giant Earth Leveller. It was powered by two Detroit Diesel 8V-71N engines, developing 575bhp between them. The machine's 6.1m- (20ft-) wide mouldboard weighed 4 tons, which gives some idea of its capacity to make the soil behave itself. RayGo also offered a 7.3m (24ft) blade as an option, but the standard issue version could level 3058 cubic metres (4000 cubic yards) of dirt in an hour.

The RayGo Giant may have been a competent machine, but it attracted no customers. The same was true of a motor grader offered at around the same time by the Oklahoma-based CMI Corporation. This was known as the Autoblade, which was a hydrostatic-drive vehicle. Like the RayGo, the Autoblade was double-articulated and fitted with a 5.5m (18ft) blade and two independent twin-axle tractor units that summoned up 375bhp. The main benefit of the Autoblade was that its central cab position could be swivelled according to the direction of travel.

Motor graders weren't completely a US phenomenon. A Canadian firm based in Ontario and called the Dominion Road Machinery Company – trading as Champion by the late 1970s – had its origins in the road-construction industry of the late 1800s and early 1900s. Dominion introduced its first motor grader, the Champion 80T, in 1975, and two years later changed its name to suit, to the Champion Road Machinery Company. The official launch of Champion's next project took place at the American Mining Congress Show at Las Vegas in 1978, and this was the Champion 100T Big Mudder. This stocky brute even had a walkway along the top of its box-section chassis arm, and boasted height of 5.5m (18ft) to the top of the cab. Fitted with a 7.3m- (24ft-) wide mouldboard blade, the Big Mudder was good for 650bhp from its single Detroit Diesel 16V-71T engine, and ought to have carried all before it. However, it was perceived in the industry as being just too big for its own good, and 10 years on, Champion sold the 100T line to the Minnesota-based Dom-Ex Incorporated mining company which ran a few other Champion graders at its sites. It was then marketed as the Dom-Ex 100T, with no conspicuous success. The Big Mudder was active in the 1990s at the Syncrude oil sand deposits in northern Alberta. Meanwhile, Champion itself was taken over by AB Volvo in 1997.

Another well-respected construction industry manufacturer, Orenstein & Koppel – better known as O&K – of Dortmund, Germany, came out with the G350 grader in 1979. It was of a more manageable size than the foregoing leviathans, wielding a 4.8m (16ft) blade and developing 300bhp. It was subsequently available with a broader 5.8m (19ft) blade, but this wasn't enough to combat the affects of the early-1980s recession, and the G350 was never sold in quantity. In fact, only 34 units had been made by the time it went out of production in 1991.

Sometimes manufacturers are caught out by extraordinary circumstances that cause orders or commissions to fall through, rather than simply making a

Hydraulic rams orientate the motor grader's blade at the required angle. Extreme blade reach means lots of side thrust, which is countered by placing a front wheel up on the bank to apply more power to the blade, while resisting the side thrust with rear steer.

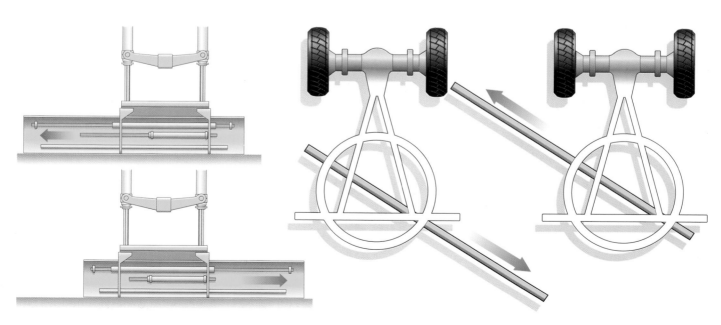

MOTOR GRADERS

bad product or being overtaken by economic circumstances. The ACCO dozer was a case in point, and the same was true of the company's gargantuan ACCO Grader. Built in 1981 when Venetian contractor Umberto Acco appeared to have a contract with Libya to supply earth-moving equipment, the ACCO Grader retained the title of largest motor grader ever. Weighing around 200 tons, it's no surprise to find the ACCO grader shod with twin wheels and tyres all round. There were only ever two ACCO graders: Acco made an experimental prototype and another more refined machine. Power units were two Caterpillar diesel engines, with a 650bhp eight-cylinder unit mounted up front driving the front axle, and a 900bhp 12-cylinder engine in the customary location driving two axles at the rear. The blade measured 10m (33ft) wide, and it could be fitted with additional 1.5m (5ft) wing-blade extensions.

The ACCO commissions were stymied by a trade embargo against Libya that prohibited commissions such as this from being fulfilled. The ACCO Grader was involved in beach grading for the town of Bibione on the Italian Adriatic coast, but spent much of its time sitting idle at the company's headquarters at Portogruaro near Venice, while the prototype was kept in store.

LOOKING GOOD

Often if something looks right, it is right, and so it was with most Caterpillar products. The state-of-the-art grader in the 1990s was almost certainly the Cat 24H. The company began developing it as the 18H in 1987, and by 1995 were ready to start trials with the prototype 24H at

AVELING-BARFORD

Make: *Aveling-Barford*
Model: *ASG13*
Manufactured: *1999*
Engine: *5.9-litre 6-cylinder Cummins turbodiesel*
Power output: *160bhp*
Transmission: *Clark, 8f/8r, (4 low, 4 high range)*
Blade width: *3.6m (12ft)*

Grading a Welsh Forestry Commission haul road is this all-wheel steer Aveling-Barford ASG13, showing typical rear-mounted engine and double-axle four-wheel layout, with single-member chassis leg ahead of the cab, the blade and circle assembly below, and axle frame up front.

Gillette, Wyoming, in the Powder River Basin, specifically in the AMAX Eagle Butte Coal Mine. The Cat 24H was fitted with a 7.3m- (24ft-) wide blade that was 1.6m (3ft 6in) tall. The whole machine was 15.8m (51ft 10in) long, and weighed in at a little over 60 tons. It was powered by the 12-cylinder Cat 3412E diesel engine, running up to 500bhp. Trials went well, and the 24H grader was released on to the market in 1996. Prime customers were large-scale mining operations and road-building contractors, for what was the largest motor grader in series production.

While the motor scraper was a basic, albeit extremely effective machine, the grader was rather more sophisticated, a reflection of the job it was created to do. Without question, both machines were an invaluable asset to the construction and mining industry.

WHEELED LOADERS AND DOZERS

These vehicles are related to tracked blade dozers by virtue of function rather than format, as they are considerably different in appearance. The average wheeled loader has four massive wheels and tyres at each corner, a cab that's accessed by the kind of staircase found on a dump truck, and a huge hydraulic shovel on the front. Much of the bulk of the engine is behind the rear axle, acting as a counterweight to the bucket shovel or blade. Having this particular piece of kit establishes the wheeled loader and wheeled dozer as a link between excavators and bulldozers, which sums up their role quite nicely: depending on the shape of their frontal attachments, they can move mountains of earth and, having done so, shovel it into waiting dump trucks.

Big rubber-tyre-shod wheeled loaders and tractors are relative newcomers on the earthmoving scene, dating back only to 1939. This Case 921C may not be the biggest in the business, a position occupied by LeTourneau, but it is an effective machine and the product of an established tractor company, which became part of the Fiat conglomerate in 1998.

WHEELED LOADERS AND DOZERS

Some years ago, these vehicles were known as tractor shovels, and production of the first one is credited to Frank G. Hough in 1939. In size and appearance it was more akin to a forklift truck with a hydraulically-operated bucket on the front, and could cope with 0.25 cubic metres (0.33 cubic yards) of soil. Hough continued to develop the vehicle type, introducing four-wheel drive with his 1.15-cubic metre (1.5-cubic yard) capacity HM in 1947, progressing to the 3-cubic metre (4-cubic yard) capacity HW model in 1950. The latter was the first wheeled loader to feature planetary axles, power-shift transmission, and torque converter, all of which have come to be considered mandatory systems on such machines.

By 1952, the Frank G. Hough Company was an attractive prospect for a big-time manufacturer, and when the firm was taken over by International Harvester, Hough's products reached a wider audience. The following year, a rival manufacturer called Mixermobile Incorporated launched a new model endowed with a radically new technical feature. Known as the Scoopmobile LD5, it was an articulated four-wheel drive swivel-steer wheeled loader. Teething troubles ensured a cool initial reception however, and it wasn't until the 1960s that bigger firms like Euclid, Caterpillar and Hough took up the concept. The idea was that the front half of the machine, incorporating the shovel, blade and hydraulics, was steered, while the powertrain was in the back section. The cab

International Hough introduced the 1.14-cubic metre (1.5-cubic yard) capacity H50B Payloader in the UK and USA in 1966. It was rear-steer 4WD, but unlike most wheeled loaders, it had a rigid frame.

was sited atop the swivel point, and on some makes it rotated with the steering and in others it remained static.

The first serious articulated-steering wheeled loader was the 300bhp Caterpillar 988, which appeared in 1963, some four years after the company first broke into the wheeled loader market with its 944A model. The 988's bucket capacity was 4.6 cubic metres (6 cubic yards) or about 1.5 tons. The formula was pretty much spot-on, as the 988 in an evolved format remained at the turn of the millennium.

In 1964, International Hough released its Cummins-powered 421bhp H-400 PAY Loader, which was an articulated loader with a 7.6-cubic metre (10-cubic yard) capacity, closely based on the company's existing D500 PAY Dozer design. Then the following year, Scoopmobile made a determined come-back with the Swivel-Steer

WHEELED LOADERS AND DOZERS

Model 1200. Despite a healthy 7.6-cubic metre (10-cubic yard) capacity bucket and 500bhp Cummins powertrain, the Scoopmobile was beset by engineering problems and couldn't be counted a huge success. The concept was attractive enough to WABCO, however, who bought up the Scoopmobile line in 1968. Having moderated the machine's shovel capacity and engine capability in a bid for better reliability, WABCO introduced its own Model 1200 in 1970, but it failed to catch the wave of new wheeled loader designs then being introduced.

TYRE DOUBT

Another firm to make its presence felt in the wheeled loader market was the Clark Equipment Company, which produced its first vehicles back in 1954. Eleven years on, in 1965, it came out with the hydraulically operated, mechanical-drive Michigan 475A, rated at 9.1-cubic metre (12-cubic yard) capacity. Powered by a 550bhp Cummins diesel, in Series III guise, the 475A was the biggest wheeled loader available then. Clark went on to produce an even larger machine, the 675 prototype with a truly impressive 18.3-cubic metre (24-cubic yard) bucket capacity – that is, 32 tons – powered along by a couple of V16 Detroit Diesel engines developing 1144bhp. Its effectiveness was hampered by inadequate tyres – Goodyear 36.00 x 51,58PRs – but these were all that were available at the time. Another tyre manufacturer, Michelin, came up with the 50.5 x 51 XRD-2 (L-5) banana-tread tyres specially for use on the 675, but these proved to be little better than the original Goodyears. It wasn't until Goodyear produced its huge 67-51SXT 54PR (L-5) covers in 1976 that development of the Michigan 675 model began to really make progress. These were the biggest tyres in existence at the time, and Clark gradually fitted them to all its customers' 675s, and all wheeled loaders manufactured post-1976 were equipped with the new 1.7m (67in) rubber. The Michigan 675 production units in Lot-3 format were powered by dual Cummins VTA-1710-C700 engines developing 1316bhp, and had a large oil cooler positioned above the rear wheel arch. With twin hydraulic rams to propel the shovel, they looked like giant grasshoppers on steroids. They were hardly prolific though, with only 18 units made up to 1988 when production ceased. The last two were actually old ones that had been rebuilt to 675C specifications and these machines were acquired by the Nerco Spring Creek Coal Mine in Montana, along with all available spares.

Truck maker Dart had experimented with rigid-chassis frame wheeled loaders in the early 1960s, and finally offered an articulated version, designated the KW-Dart D600, in 1965. This model was equipped with a 11.5-cubic metre (15-cubic yard) bucket shovel

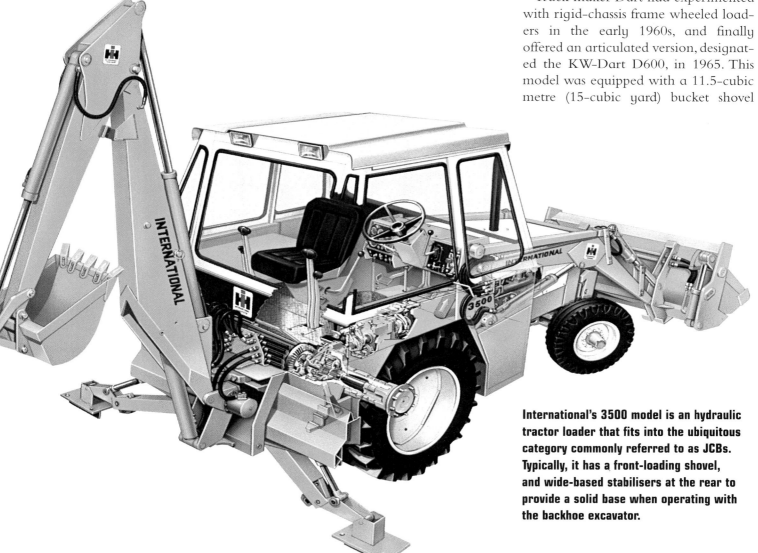

International's 3500 model is an hydraulic tractor loader that fits into the ubiquitous category commonly referred to as JCBs. Typically, it has a front-loading shovel, and wide-based stabilisers at the rear to provide a solid base when operating with the backhoe excavator.

and ran with a 600bhp Cummins powertrain. The cab was mounted to one side, which not only gave the driver a better view of his task, but made the vehicle more readily identifiable, in spite of its orange paint finish. The D600 was the first of a long lineage of Dart wheel loaders, beginning with the D600 Series B model from 1973, the 600C of 1978, and the 600E coal ejector bucket loader that came out in 1982. In all, some 400 units were made of the 600 series, and engine options were the Cummins KT-38C or the Detroit Diesel 12V 149TB. Overall capacity didn't alter much, rising to 12.2 cubic metres (16 cubic yards) for the 600C model, and 17.6 cubic metres (23 cubic yards) with a coal ejector bucket, by 1982. Production dwindled during the following decade.

Dart dallied briefly with diesel-electric powertrains, too. In 1974, the company brought out its DE620 wheeled loader to go head to head with Marathon LeTourneau's contemporary L-700A model. The 650bhp DE620 was rated at 11.4-cubic metre (15-cubic yard) capacity, and there was also a proposal for a vehicle to be designated the DE900 running with a 1250bhp Detroit Diesel motor and hoisting a 16-cubic metre (21-cubic yard) bucket. Like the LeTourneau model, the Dart DE620 used a diesel-electric powertrain, in this case with a single electric traction motor connected to each axle housing. The layout proposed for the DE900 located each traction wheel motor in the hub assemblies of each driving wheel, as did the LeTourneau layout. The bigger unit didn't even make it into prototype stage, and the DE620 was only built in minute numbers: 12 units, to be precise.

UNIT RIG OWNERSHIP

Although some units of the Dart 600 series were made and marketed under licence in Canada as Sicards in the late 1960s, by this time, the company was under the ownership of Unit Rig, and the final Dart 600 wheeled loader left the company's Tulsa factory in 1995.

Going back to the mid-1960s, probably the most significant model to emerge was Caterpillar's 992. This machine went on to become top-seller in the articulated 7.6–11.4-cubic metre (10–15-cubic yard) wheeled loader market segment, and the series endured into the late 1990s. Field tests with pre-production units began in 1965, and the 992 model went on sale in 1968. Powered by a 550bhp Cat engine, and originally fitted with a 7.6-cubic metre (10-cubic yard) capacity shovel, the 992 was upgraded in 1973 as the 992B and in 1977 as the 992C. The latter's increased performance was derived from its 650bhp Cat engine and redesigned hydraulic lifting arms, which enabled it to accommodate a 9.5-cubic metre (12.5-cubic yard) shovel. For some 15 years, this proved quite satisfactory, and only in 1992 did the model evolve into the 710bhp 992D, fitted with a 10.7-cubic metre (14-cubic yard) bucket shovel. The next upgrade was in 1996, with the introduction of the 992G, measuring 16.1m (53ft) long and rated at 800bhp with a 12.2-cubic metre (16-cubic yard) bucket capacity. If that doesn't sound particularly innovative, the revised model, featuring a new front-lifting arm, consisted of an individual cast-steel box section unit instead of the previous twin-boom format. This was a more tidy arrangement, and offered the driver increased visibility when dumping the contents of the bucket.

International Hough came up with an enormous 13.7-cubic metre (18-cubic yard) prototype that represented a quantum leap in loader carrying capacity in 1971. This was the 580 PAY Loader, and it was some six years in development prior to its launch in 1978 when capacity had risen to a substantial 16.8 cubic metres (22 cubic yards). Powertrain was a solitary 1075bhp Detroit Diesel unit, good enough to weather the economic turbulence of the early 1980s which saw International sell off its construction equipment arm to Dresser Industries in 1982, only for Dresser and Komatsu to

Doing his homework perched on the platform of a Caterpillar 994D, the surveyor with his lap-top is some 3.5m (12ft) off the ground, demonstrating the sheer scale of these wheeled loaders.

WHEELED LOADERS AND DOZERS

Caterpillar's giant 994D snuggles up to a rigid-rear dump truck to deposit its load. Machines like this can fill a big dumper in just four passes, and the big Cat displays impressive loadover height and reach for such a large machine.

amalgamate in the KDC joint venture company in 1988. By 1991 the 580 had metamorphosed into the longer wheelbase Haulpak Dresser 4000 wheeled loader, boasting a 1280bhp power rating from its Detroit Diesel 12V-19TI DDEC engine, or 1250bhp from the Cummins KTTA-38-C unit, and a huge bucket capacity of 18.3 cubic metres (24 cubic yards). However, sales proved insufficient to maintain production, and the model reached the end of the line in 1995, leaving a gap of three years before Komatsu returned to this sector with the 18.3-cubic metre (24-cubic yard) capacity WA1200-1 model.

MONSTER DEVICE

Following in the tyre-tracks of the Clark Michigan 675, the Japanese Surface Mining Equipment for Coal Technology Research Association – or, mercifully, SMEC for short – came up with a monster twin-engined device in 1986. SMEC was a Japanese government quango consisting largely of construction equipment manufacturers who sought to develop a big wheeled loader and other coalmining exploration equipment. Despite the presence of Komatsu in the marketplace, it was thought that a wheeled loader of gigantic proportions could be built were the costs to be met by a consortium. The resulting vehicle was the SMEC 180t Super Wheeled Loader, built almost entirely by Kawasaki Heavy Industries.

The SMEC Kawasaki was shod with 1.7m (67in) Bridgestone tyres, and the twin-engine concept was revived with a pair of 12-cylinder Cummins VTA-28C diesels, developing a stout 1340bhp. The capacity rating of its bucket shovel was 34 tons, which translated as 19.1 cubic metres (25 cubic yards). The prototype underwent shakedown tests at Okumura Gumi Civil Engineering company's quarries on Nishijima Island in 1987, relocating

---

CAT 994D

Make: *Caterpillar*
Model: *994D*
Manufactured: *1990*
Engine: *16-cylinder Cat 3516 turbodiesel*
Transmission: *Cat planetary power-shift*
Power output: *1250bhp*
Bucket capacity: *17.6 cubic metres (23 cubic yards)*

to Australia for further evaluation. Although everything went according to plan, the SMEC 180t failed to capture the imagination of the world's mining industry, and the prototype remained just that, albeit lifting the title of world's biggest wheeled loader for a time.

Continuing in a Japanese vein, Komatsu was involved with developing its own WA800 wheeled loader at the time when Kawasaki was building the SMEC vehicle. Targeting Caterpillar's 992C customers, and indeed anybody else in that segment, the 10.5-cubic metre (13.7-cubic yard) capacity WZ800-1 was unveiled in 1986. It used an in-house 789bhp Komatsu SA12V140Z-1 diesel engine, with a second-generation model coming out in 1988 called the WA800-2. A further variant was introduced a decade on in 1998 as the WA800-3 model, which attained 800bhp. By this time though, it had been largely superseded by the WA900-1. Debuting at the Las Vegas MINExpo in 1996, the WA900-1's Komatsu V12 engine pushed out 850bhp to provide the energy to drive its 13-cubic metre (17-cubic yard) bucket.

The earth-moving industry is a fairly tightly-knit world, and rumours and stories often of great intimacy abound. Thus, when Caterpillar released its range-topping 994 model in 1990, many peoples' suspicions were justified, as a large wheeled loader had been anticipated for some time. The mechanical-drive Cat 994 was fitted with a 17.6-cubic metre (23-cubic yard) rock bucket as standard, with a 34.4-cubic metre (45-cubic yard) coal-loader as an option. It was designed to be used in conjunction with face shovels and to load dump trucks rated at 168 tons, which it could fill with just four bucketsful. A high-lift option was available for accessing higher-sided vehicle hoppers. The chosen engine was the V16 Cat 3516, developing 1250bhp, transmitted via the Cat planetary gear power shift. This was a time when the heavy-plant tyre manufacturers lagged behind vehicle technology, just as can happen in motor racing when mechanical systems leap forward. Initially running on compromised tyre sizes, the 994 had to wait until suitable rubber-wear became available, and these turned out to be new 53.5/85-57 and 55.5/80-57 covers. It wasn't long before the model itself was revamped, with the similarly rated 994D model appearing in 1999, powered by the Cat 3516B motor.

DIESEL FITTER

It was feasible to imagine that wheeled loader and dozer manufacturers would opt to try diesel-electric power systems. After all, they had worked well enough

WHEELED LOADERS AND DOZERS

Hard at work shifting rubble in a quarry is this Case 921C. In 1999 the tractor company amalgamated with New Holland, bringing it under the umbrella of the Italian Fiat conglomerate.

in rear-dump haulers. However, only one maker rose to the challenge, and that was Robert Gilmour LeTourneau. From his Longview, Texas base, LeTourneau set out to produce a viable electric-powered wheel system, funded by the sale of much of his heavy-plant business to WABCO back in 1953. When he was free of the five-year moratorium that banned him from competing with WABCO, LeTourneau unveiled a raft of new designs, utilising his diesel-electric drive system and traction wheel motors. The scrapers came first, followed closely by wheeled loaders such as the SL-10 Short Lever Shovel that debuted in 1960. Somewhat unusually, this was a three-wheeler model, fitted with a 7.6-cubic metre (10-cubic yard) bucket. Largest and most powerful of R.G. LeTourneau's early diesel-electric drive Pacemaker tractors was the K-205 Series, released in 1961. This extraordinary looking machine was a five-wheeler – the fifth was right at the rear of the vehicle – and was powered by a pair of Cummins V12 diesel engines producing 1100bhp and mounted in solitary splendour on the rear platform. Hub-mounted electric traction motors drove all five wheels. Its overall length was 15.3m (50ft 3in), and the shovel blade measured 6.1m (20ft) wide. But more conventional four-wheel vehicles with articulated-chassis steering soon arrived, including the 11.4-cubic metre (15-cubic yard) capacity SL-30 of 1964 and the prototype for the 14.5-cubic metre (19-cubic yard) SL-40. The latter was the biggest traditional wheeled loader in LeTourneau's catalogue, and it was certainly the weirdest looking. The cab was located over the front left-hand wheelarch, and the twin Detroit Diesel 12V-71N motors sat in a triangular-profile raft at the rear, behind the back axle, punting out some 950bhp. The SL-40 made its debut at that popular venue, the American Mining Congress Show at Las Vegas, in 1965.

ELECTRIC DRIVE

LeTourneau's T-600 series tractor was among the company's largest electric-drive models, using the obsolescent rack-and-pinion system to control the blade. It was based on an articulated steering chassis and was originally designed in 1967 with a push-block as a tractor for scraper loading applications. A 600bhp Detroit Diesel 16V-71N engine powered it, with four-wheel drive provided by the electric-traction motors housed in each wheel, and a 4.8m (16ft) wide dozer blade could be specified. Shortly afterwards, LeTourneau released the smaller 5.7-cubic metre (7.5-cubic yard) SL-15 and 7.6-cubic metre (10-cubic yard) SL-20. They all had electric-drive and rack-and-pinion gear assemblies to perform the motive functions, which included steering, raising and lowering the shovel boom, and emptying the bucket.

What, no hydraulics then? Quite so. LeTourneau preferred to rely on electricity, but reluctantly he bowed to the demands of the marketplace and replaced the antiquated electro-mechanical systems with modern hydraulics. First up with hydraulic controls in 1968 was the prototype known as the XL-1 LeTric-Loader, and this was to be the blueprint for all future LeTourneau wheeled loaders. The logo on the vehicle's flanks made a play on the 'Le' part of LeTourneau. It was a massive, two-section articulated chassis, with no protection whatsoever for the driver. Like some of the big rigid dump trucks made by Dart and WABCO, it had a diesel engine that generated the power for the electric motors in all four wheel hubs. In this case, that was a 700bhp V16 Detroit Diesel 16V-71T-N75 engine. Field tests began on the 11.4-cubic metre (15-cubic yard) prototype L-700 – restyled the Letro-Loader – in 1969. A smaller model rated at 7.6 cubic metres (10 cubic yards) and designated the L-500 also appeared in prototype form the same year, although this was suspended when it became clear that it was more prudent to fund development of the larger vehicle.

A corporate sea change was in the air, as R.G. LeTourneau died in June 1969, and over a year later the business

WHEELED LOADERS AND DOZERS

Its paint work is so pristine that this Kawasaki 90 ZIV clearly hasn't yet turned a wheel in anger. With bucket raised, it demonstrates the full vertical extent of its hydraulic lift capability. Its front differential housing is also obvious.

was acquired by the Marathon Manufacturing Company. By 1974, it had come to be identified as Marathon LeTourneau. This had little bearing on the success of the L-700 Letro-Loader, which accounted for sales of 76 units up to 1975, when it was superseded by the L-800 model which had been developed on the same platform. Although the L-800's shovel capacity was no greater than its predecessor's, it ran with the more powerful 750bhp Detroit Diesel, or alternatively, the 675bhp Cummins unit. Between 1975 to 1983, Marathon LeTourneau sold 193 units of the L-800 wheeled loader which indicated that mining companies recognised the potential of diesel-electric power.

The 525bhp Marathon LeTourneau L-600 of 1977 was another venture into the 7.6-cubic metre (10-cubic yard) capacity segment, but it seemed that the company was never destined to reap success at this weight. It wasn't down to any particular inadequacies in the vehicle's specification, just that it was immediately confronted by Caterpillar's revitalised 992C model, and this was the one the buyers went for. A mere 26 units of the L-600 loaders were ordered.

WHEELED LOADERS AND DOZERS

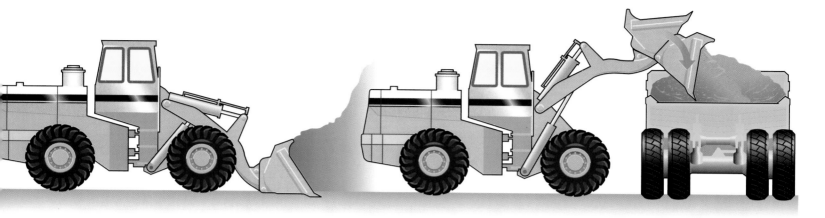

Size-wise, the enormous L-1200 Letro-Loader was a different proposition altogether, making its debut at the 1978 American Mining Congress Show at Las Vegas. A perfectly reasonable prospect for a voracious mining industry when conceived – boasting a 16.8-cubic metre (22-cubic yard) bucket and a 1200bhp Detroit Diesel 12V-149TI power unit – the L-1200 Letro-Loader never realised its potential. It was sunk by the world recession that held sway in the early 1980s, and only three prototypes and eight production units were made. Six of these found work in a coalmine in Colombia, South America, and one went to the Boliden Minerals copper mine at Gallivare in Sweden. Happily for Marathon LeTourneau, its other heavy-plant machinery kept it buoyant, and the reasonably successful L-800 series was superseded in 1983 by the bigger L-1000, which carried a 13-cubic metre (17-cubic yard) bucket shovel and 900bhp power unit.

Marathon LeTourneau had the 16.8-cubic metre (22-cubic yard) capacity segment covered with its 1050bhp L-1100, launched at the 1986 American Mining Congress Show. Although on paper its carrying capacity statistics were identical to the outgoing L-1200, the new vehicle was a wholly new design. Four years on, the company came out with its masterpiece wheeled loader, the L-1400 Letro-Loader which, uniquely at this time, was fitted with a 21.4-cubic metre (28-cubic yard) rock bucket. Also available was a 32.8-cubic metre (43-cubic yard) coal-loader combo-bucket for mining applications. It was offered with either Detroit Diesel 16V-149TI DDECIII or Cummins K-1800E power units that developed 1800bhp. The same choice of engines in 2000bhp format could be specified in the L-1800 Letro-Loader of 1993, which again surpassed in size all previous wheeled loaders. Its standard rock bucket capacity was 25.2 cubic metres (33 cubic yards), with 42 cubic metres (55 cubic yards) available later,

Wheeled loaders work with hydraulically-operated buckets. Pick-up is a three-part procedure. The machine first scoops a bucketful of material, angles the bucket by 45 degrees to contain it, then in the third phase, hoists it aloft ready to deposit it.

which was sufficiently capacious to tackle work that was previously the exclusive province of cable or hydraulic mining shovels.

ROWAN TAKEOVER

The following year, the Marathon LeTourneau Company was taken over by the oil exploration firm Rowan Incorporated, which had been among LeTourneau's biggest and longest-standing clients for oil-rig platforms. As part of the makeover that all newly acquired companies get, for better or worse, the Marathon identity was abandoned, and the original 'L' script featured in the revised LeTourneau paint scheme and logo. One of the first machines to feature in the new corporate livery was the diesel-electric L-1350, a 19.8-cubic metre (26-cubic yard) capacity wheeled loader that hit the dirt in 1999. The Rowan takeover inspired the production of no fewer than 570 Letro-Loaders of various capacities between 1998 and 1999.

Want to know what one of these giant wheeled loaders would set you back? LeTourneau's biggest offering, the L-1800, costs £1.8m ($3m), takes a

Undaunted by the size of the boulders it's faced with, the Komatsu WA600 gets stuck in. This model shares the same configuration as its bigger sister, the WA800/900.

WHEELED LOADERS AND DOZERS

Top of JCB's 17-model range of wheeled loaders, the 456B features a Z-bar design loader end, dual-mode hydraulics, load-sensing piston pumps, fully automatic powershift transmission and low emissions engine, single-lever servo control and electronic monitoring system.

JCB 456B

Make: *JCB*
Model: *456B ZX*
Manufactured: *1998*
Engine: *Cummins C-series turbo-diesel*
Power output: *153bhp*
Operating weight: *18 tons*

WHEELED LOADERS AND DOZERS

week to assemble and has a cab filled with creature comforts to cosset the driver – like air-con, stereo hi-fi and suspension seating. The teeth on the front of its massive rock bucket – big enough to accommodate a family saloon – get replaced every six to eight shifts. Its tyres – some 4m (13ft) tall – each cost as much as three cars, and take an hour to change with the aid of cranes.

Other global manufacturers that produced wheeled loaders during the 1990s included Atlas, Case, Daewoo, Halla, Hyundai, JCB, Liebherr, O&K, Samsung and Volvo. Typical of the Halla Heavy Industries range was the HA380 model, which used a 216bhp turbocharged Cummins LTYA-10C engine and had a 3.7-cubic metre (4.9-cubic yard) bucket capacity. Its oscillating rear axle allowed sufficient axle articulation for it to be operated on particularly rough terrain. O&K's Dortmund-built range included the entry-level 1.4-cubic metre (1.9-cubic yard) L15 rising to the range-topping L45 offering a 3.7-cubic metre (4.9-cubic yard) capacity. This model was driven by the 177bhp Cummins LT-10C, and its 37km/h (23 miles/h) forward and reverse-speed capability demonstrated the marked difference between wheeled and crawler loaders.

Volvo's C-generation of wheeled loaders consisted of seven different models, starting with the L50C and ending with the L330C. These vehicles all used steering-wheel control, and an articulated front frame with fully floating axles carried the bucket arm and hydraulics. The mid-range models provided a Comfort Drive Control facility, which permitted the driver to steer and change gear with his left hand and operate the loader system with his right. The big L33C was powered by Volvo's own straight-six 330bhp TD 164 KAE turbodiesel engine and had a 6.3-cubic metre (8.3-cubic yard) capacity. Like all modern wheeled loaders, the cab was air-conditioned, cushioned against vibration, and fitted with a seat suspension. Further specialised models in Volvo's repertoire were the L180C Wheeled Timber Lifter, fitted with an enormous grab on the same principle as those you see in seaside amusement arcades fishing for

WHEELED LOADERS AND DOZERS

Like all big wheeled loaders, the Komatsu WA900 has a healthy appetite for the rough stuff. Introduced in 1996, the WA900 shared the same external dimensions as the WA800-2, but was fitted with a more powerful 853bhp Komatsu SA12V140Z-1 turbodiesel.

trinkets, only this one picked up tree trunks. For use on landfill sites, Volvo offered the L180C Wheeled Loader Compactor. This was fitted with body armour on its sides and underneath as standard, and a set of wheels of varying styles, all reminiscent of medieval battle axeheads, was offered instead of pneumatic tyres that would easily puncture in harsh operating environments.

WHEELED DOZERS

Shod with rubber tyres, the wheeled dozer combined the functions of the more agile articulated-chassis wheeled loader with the more pugilistic heap-shifting qualities of the crawler dozer or blade. The only thing to notice is that wheeled-dozer blades tended to be less elaborate than crawler-dozer blades, which were more specialised. Being wheeled, they are also more manoeuvrable than the crawler variety. The history of the wheeled dozer is not quite as long as the wheeled loader, but not far off it. It is generally credited to the enormously innovative R.G. LeTourneau, who came out with the wonderfully named Tournadozer, in 1946. This was simply a rigid-chassis rubber-tyred tractor that used a pneumatic clutch system to lock the driving wheels on either side to steer it. The Tournadozer C1 was introduced in 1947, and further experimental versions followed in the next couple of years, including the Super Model A tractor, unveiled at the 1948 Chicago Road Show and powered by a 600bhp Packard marine engine, which was more than the chassis could handle.

LeTourneau's electric-drive wheeled dozers matched the company's massive scraper plant in terms of scale, and the biggest of its late-1950s models was the Pacemaker Series K-205 Tractor. This five-wheeled beast had an electric motor in each wheel hub, with its main engines being a trio of 350bhp V12 Cummins diesels mounted at the rear and developing 1050bhp. The business end was a 6.1m- (20ft-) wide dozer blade. To say that it looked unconventional was to be guilty of gross understatement, and this opinion appears to have been held by much of the mining industry. However, LeTourneau persevered, and in 1965 came up with the LeTourneau K-54 Dozer. Not only did it have electric traction wheels, but it also incorporated the loader's articulated chassis steering. The 400bhp K-54 was followed in 1966 by the 275bhp T-300-A, and a year later came the 500bhp T-600-A. The XT-1 project of 1969 was based on the ill-fated L-500 Letro-Loader and evolved into the 475bhp LeTourneau D-450B, suitably equipped with hydraulic controls for all blade operations. The economic circumstances that did for the L-500 loader also sank the D-450-B tractor dozer and just the single prototype was produced. The last gasp of the diesel-electric traction-wheeled dozer was the D-800 Letro-Dozer, unveiled by Marathon LeTourneau at the American Mining Congress Show in 1978. Naturally derived from the L-800 wheeled loader chassis, the D-800

WHEELED LOADERS AND DOZERS

was powered by a rear-mounted 800bhp diesel engine. But just as the L-1200 loader was doomed by the early 1980s recession, so the same fate befell the D-800.

Only one other manufacturer came to the market with a machine that ran with an electric-drive system, although it turned out to be a one-off. In 1963,

Most wheeled loaders have articulated steering and four-wheel drive. With a breakout force that can lift at least 20 tons, the machine raises its laden bucket to full loadover height and tips it into the waiting dumper.

the Western Contracting Corporation launched its Western 2000 tractor, designed by C.W. Jones Engineering Company of Los Angeles, who, in 1958, had also designed the monstrous Western 80 articulated dump truck known as the Eucnik. The Western 2000 tractor unit was constructed in just 10 months by the Parkville, Missouri-based Intercontinental Engineering Manufacturing Corporation. In fact it wasn't a wheeled dozer, but was a tractor intended for scraper loading applications. Power came from a colossal 1650bhp General Motors 16-27EA marinised diesel

Liebherr's medium-duty 544 model delivers a bucketful of overburden into a waiting dumper. Like the majority of modern loaders, the Mercedes-Benz-powered Liebherr is effectively hinged in the middle, aft of the cab, and the articulation thus provided makes for good mobility and a tight turning circle.

engine, transmitted to General Electric traction motors in the ends of the axles, and an auxiliary 200bhp GM diesel unit powered the AC generator. The Western 2000 was shod with specially made Goodyear tyres, 3m (10ft) in diameter. It remained unique as the largest push-dozer ever made.

WHEELED LOADERS AND DOZERS

The Caterpillar 834B was introduced in 1964 as the Cat 834. The wheeled dozer combined the functions of the more agile articulated-chassis wheeled loader with the heap-shifting qualities of the crawler dozer or blade.

BEST OF THE REST

Other manufacturers who built big conventional diesel-powered wheeled dozers included Clark Michigan, who offered the Michigan Model 480 as a scraper tractor from 1958. This used the 500bhp Cummins VT-12 engine, with the Detroit Diesel 16V-71N unit as an option, and it was available with push-block for scraper loading or 4.2m (14ft) dozer blade. Between 1958 and 1966, only 90 units were produced.

International Hough came out with the large pivot-steer pusher-dozer identified as the Hough D-500 PAY Dozer in 1959, and available from 1961. It was fitted with a detuned 600bhp Cummins VT-12-700-CI diesel engine, and subsequently used a 550bhp Detroit Diesel 16V-71N unit. The D-500 came with a push-block for scraper loading or a 4m (13ft 5in) dozer blade. The company's H-400 PAY Loader, which had a 7.6-cubic metre (10-cubic yard) capacity was closely based on the D-500 wheeled dozer. Other wheeled dozers included the Allis-Chalmers 555 of 1963, the FWD-Wagner W1-30 pusher-dozer, the RayGo Grizzly 150 that appeared in 1971, and Melroe's M880 Multi-Wheel Dozer dating from 1976. One remarkable vehicle that came out in 1970 was the Peerless Manufacturing Company's VCON V-250 prototype diesel-electric drive wheel tractor. Intended to push-load scrapers, the Dallas-built VCON V-250 spent its first 1000 hours working at a uranium mine, after which the push-pad was swapped for a 6.7m- (22ft-) wide dozer blade. It went on to operate at a number of North American mines, including the oil tar sands of Alberta.

The VCON (Vehicle Constructors) division of Peerless was acquired by the Marion Power Shovel Company in 1974, and the V-250 project evolved into the V-220. This imposing 1320bhp diesel-electric wheeled dozer was in the metal by 1975, and tipped the scales at 150 tons, making it the biggest vehicle of its type in the world. Motive power was provided by a Detroit Diesel 16V-

CAT 834B

Make: *Caterpillar*
Model: *834B*
Manufactured *1990*
Engine: *Cat 3508B EUI turbodiesel*
Power output: *800bhp*
Blade width: *6.2m (20ft 7in)*
Blade capacity: *25.2 cubic metres (33 cubic yards)*

149T engine driving four GE electric motors, with one to each wheel. It was equipped with a 7.9m- (26ft-) wide semi-U blade, and the two units made proved highly effective in land-reclamation duties. However, when Marion was taken over by Dresser Industries in 1977, the VCON dozer prototypes were put out to grass.

AUSTRALIAN MINING

There's plenty of mining activity in Australia, and Tiger Engineering Pty was well placed to exploit it. Founded in 1980, Tiger unveiled its first mechanical-drive wheeled dozer in 1981. The Tiger 690A relied heavily

WHEELED LOADERS AND DOZERS

Crawler loaders are used where traction is vital, on particularly rough construction sites or boggy ground, for example, and rigid models like this Cat 933 Hystat provide a more rugged solution than wheeled loaders. The two-model Hystat range includes the turbocharged 939C and normally-aspirated 933C.

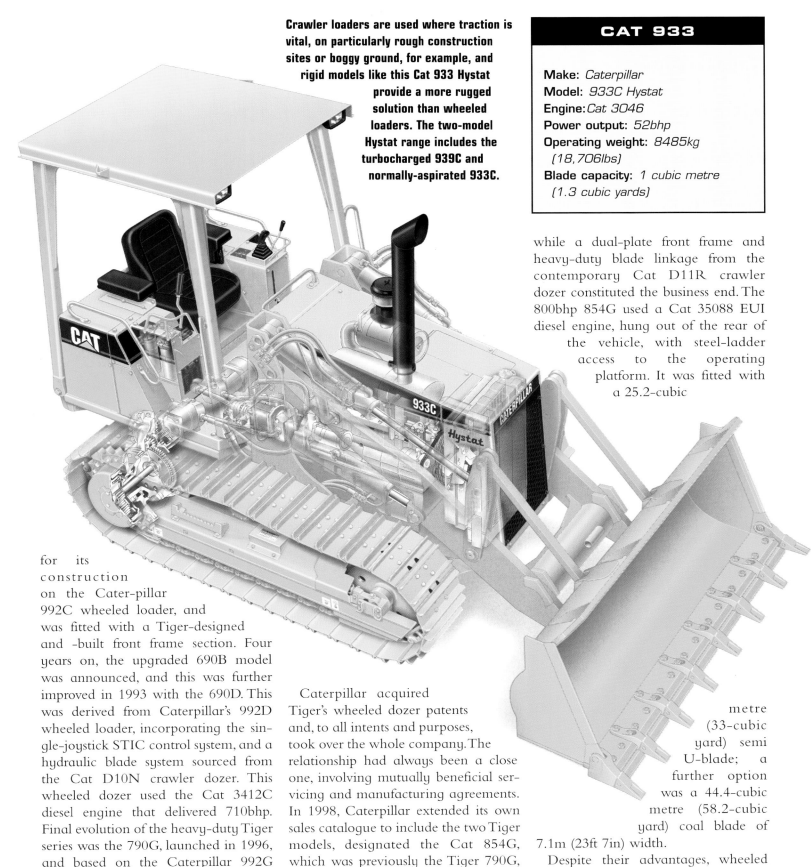

CAT 933

Make: *Caterpillar*
Model: *933C Hystat*
Engine: *Cat 3046*
Power output: *52bhp*
Operating weight: *8485kg (18,706lbs)*
Blade capacity: *1 cubic metre (1.3 cubic yards)*

for its construction on the Cater-pillar 992C wheeled loader, and was fitted with a Tiger-designed and -built front frame section. Four years on, the upgraded 690B model was announced, and this was further improved in 1993 with the 690D. This was derived from Caterpillar's 992D wheeled loader, incorporating the single-joystick STIC control system, and a hydraulic blade system sourced from the Cat D10N crawler dozer. This wheeled dozer used the Cat 3412C diesel engine that delivered 710bhp. Final evolution of the heavy-duty Tiger series was the 790G, launched in 1996, and based on the Caterpillar 992G loader. Powered by the 800bhp Cat 3508B EUI diesel engine, the 790G was the most powerful wheeled dozer that Tiger ever made.

Caterpillar acquired Tiger's wheeled dozer patents and, to all intents and purposes, took over the whole company. The relationship had always been a close one, involving mutually beneficial servicing and manufacturing agreements. In 1998, Caterpillar extended its own sales catalogue to include the two Tiger models, designated the Cat 854G, which was previously the Tiger 790G, and the Cat 844, previously the Tiger 590B. The Cat 854G's rear chassis element and powertrain was derived from the Caterpillar 992G wheeled loader, while a dual-plate front frame and heavy-duty blade linkage from the contemporary Cat D11R crawler dozer constituted the business end. The 800bhp 854G used a Cat 35088 EUI diesel engine, hung out of the rear of the vehicle, with steel-ladder access to the operating platform. It was fitted with a 25.2-cubic metre (33-cubic yard) semi U-blade; a further option was a 44.4-cubic metre (58.2-cubic yard) coal blade of 7.1m (23ft 7in) width.

Despite their advantages, wheeled dozers were not as prolific as tracked ones, although plenty of mining- and quarry operators found them eminently serviceable.

DUMP TRUCKS

Clearly, they're trucks on steroids, scaled-up Tonka Toys for giants, and decidedly cute in a chunky kind of way: that's to say, viewed from a distance, because up close, they're as large as a house. But let's get a few definitions straight before we start. The dump truck that evolved from regular construction site trucks during the 1930s is also variously known as a rear-dump hauler, a dumper, a mining truck or an off-highway truck. Dumpers usually refer to the smaller variety of off-road machine often seen on housing developments, while our subject matter here, the giant dump trucks and ultra haulers, are found in the world's remoter quarries and mining locations.

Back in the 1920s, regular lorries – they still call them wagons in the north of England – had their load-carrying beds adapted with somewhat ramshackle shielding over the cab for transporting large quantities of rock and rubble when operating in construction sites and quarries. More than likely, they started out with wooden bodies, and when these wore out, they were replaced with steel ones. Among the first to adapt was the fabled US manufacturer Mack, who fitted its 1929 AP model with special 10.7-cubic metre (14-cubic yard) dumper bodies. This was

Dump trucks evolved as a special breed of off-highway vehicle as a result of major construction projects during the 1930s, achieving their present incarnation in the 1960s. Here, in a typical scenario, a Cat 998 wheeled loader fills a Cat 771D at a quarry site.

DUMP TRUCKS

DUMP TRUCKS

International released its 32-ton capacity 100 Payhauler off-highway dumper truck in 1965. This 375bhp rigid rear-dump model drove its rear axle through six forward speeds via torque converter transmission.

in response to the specific needs of the Boulder Dam project – now known as Hoover Dam – around 1931, and they were in reality not much more than muscular versions of the AC company's legendary Bulldog model. Probably the first company to build a dedicated dump truck was the Euclid Crane and Hoist Company of Cleveland, Ohio, who brought out a trial model in 1933 known as the ZW. After tests, the ZW was productionised as the Model Z and known in the trade as the Trac-Truk. The Model Z was the precursor of all off-road dump trucks. It fast became the norm to define such vehicles not so

Terex 2366 dump trucks line up for a day on the haul road. With plants in Motherwell, Scotland and Tulsa Oklahoma, the Terex heavy equipment ranges includes six rigid and six articulated off-highway dump trucks of varying payloads.

much by axle weights or engine capacities, but by the volume of matter they could haul. Thus, the Trac-Truk's carrying capacity was identified as 5.35 cubic metres (7 cubic yards), or just over 16 tons. As payload capacities increased, this figure came to be expressed in tons.

War was the catalyst and accelerator for technological developments and as far as heavyweight dump trucks were concerned, the chief benefit was the advance in hydraulics. This meant bigger and better rams could be incorporated into the vehicles' tipping mechanism. In the UK, truck makers Foden came out with a specially designed dump truck in 1948. It was a tandem-drive 6x4 model, powered by the Foden FED6 two-stroke supercharged diesel engine. During the following decade, Foden, along with other British heavy-duty truck producers Thornycroft and Scammell, built rigid rear-dump haulers. Equally as a result of military demands, General Motors Allison automatic transmissions and torque converters were adapted for off-road use by Euclid from 1947 onwards in the

shape of twin-engined powertrains. These had been used in military tanks and were first installed for civilian use in the tractor unit of Euclid's bottom-dump 50FDT-102W of 1948. The company's 1FFD dump truck of 1949 was equipped with a pair of GM diesel engines that churned out 380bhp, which was not at all bad for the time. This vehicle featured tandem-drive via an Allison Torqmatic transmission set-up and was rated at 34 tons capacity. The two-engine transmission concept was trademarked as the Twin-Power system and it facilitated synchronised gear-shifting of the combined unit.

By spring 1951, Euclid was ready to launch its 1LLD prototype. Like a massive bricklayer's hod, it featured the broad, ribbed-steel body, with protective, heavily ribbed steel tray over the cab to stop rocks or overburden dropped from an excavator caving in the cab. This typified the dump truck layout. It was destined for the Western Contracting Corporation of Sioux City, Iowa, and put to work on the Fort Randall Dam project in South Dakota. The Euclid 1LLD was powered by two

Cummins NHRS six-cylinder diesel engines, with a combined output of 600bhp at the tandem back axles. In its final form, the tandem-drive model R-62 5LLD was powered by two GM Detroit Diesel 6-110 engines, developing a combined total of 632bhp.

Euclid continued to consolidate its position at the forefront of the industry, and among its best-known models was the diesel-electric R-170 model. Available from 1975, it was still in production in the late 1990s. Its capacity rating was originally 170 tons, which was subsequently elevated to 185 tons. The R-170 was powered by a 1519bhp 16-cylinder Cummins KTA50-C diesel as standard, and the 1492bhp Detroit Diesel engine was optional fitment. You'd be forgiven for thinking that throughout the heavy-plant industry, colour schemes were unswervingly yellow or orange. It was suggested that this makes them harder to detect if stolen, as all such vehicles look alike, at least from a distance. But Euclid trucks were normally painted Hi-Lite green for the North American market, and what was termed Volvo yellow elsewhere, although buyers were free to specify their own colour scheme. If the colour scheme was discernible under the dust and mud that characterised their operating environment, Liebherr haulers were actually finished in white.

By 1986, Euclid's R-190 dump truck – with 190 tons capacity, as the designation suggests – was available. The engineering of these vehicles is so thorough and in a sense, basic, that they endure

Unit Rig & Equipment experimented with gas turbine power in modified Lectra Haul M100s in 1965, using 1100bhp General Electric LM-100 gas turbine and 1000bhp IH Solar Saturn gas turbine engines. But despite promising tests, they were later restored to regular diesel-electric format.

for a long time, far longer than any self-respecting car maker would countenance. So it wasn't until 1996 that Euclid announced an updated version of the R-190. A facelift sounds too cosmetic a term for these macho monsters, but the changes were evident in the vehicle's revised frontal aspect. It was still available with two engine options, both rated at 1650bhp – the Cummins K1800E or the Detroit Diesel 16V-149TIB units – taking cargo capacity up to 202 tons. By this time, Euclid had combined with the Japanese Hitachi concern as Euclid-Hitachi. The squat-looking R-260 dump truck was unveiled in September 1996 at the MINExpo event at Las Vegas, Nevada. Powered by a 2390bhp Detroit Diesel/MTU S-4000 DDEC engine, the R-260's carrying capacity was 262 tons. What made it look different from rivals was the ladder platform and half-spiral staircase that flanked the vehicle's vast radiator grille. Like the majority of dump trucks, the huge air cleaners on either side give the appearance of giant twin headlights, which in reality are tiny units located high up in the radiator façade. As was customary, the driver sat high in his cab – as if on the bridge of a ship – over to the left of the vehicle.

OLD HANDS

Another time-served truck maker that came into the dump truck arena early on was Dart. This firm's history went back to 1903, and its first dump truck was the tandem-drive Model 60, designed by Ralph H. Kress and launched in 1952. It was subsequently identified as the 75TA, and had a curious, if logical layout. Each of its two 350bhp Buda Super Diesel engines was located low down on the outside chassis frame rails just behind the cab in order to obtain a lower centre of gravity. The prototype Dart 60-tonner was delivered to the Baghdad Copper Corporation in Arizona in 1953. However, it became clear from the outset that the Buda engines were not large enough to cope in a mining scenario, and unfortunately this prompted Baghdad Copper Corpora-

DUMP TRUCKS

Liebherr Mining Equipment's rear-dump trucks were fitted with Siemens AC-electric traction wheel drive systems, including the 300-ton T252 model shown here being loaded in a Spanish quarry by an excavator from the same stable, the Litronic 995 mining shovel with 26.7-cubic metre (35-cubic yard) backhoe.

tion to pull the plug on further commissions. When Dart came out with the Model 60 uprated to a 75-ton capacity in 1954 as the Model 75TA, its reputation was beyond redemption in the mining industry.

Somewhat later in its history, Dart came up with the DE930, a 120-ton capacity vehicle. It was launched in 1971. While previous Darts had been converted to run as diesel-electric packages, the DE930 was designed specifically as a diesel-electric dump truck, with the intention of being one of the company's landmark vehicles. The DE930 used a 900bhp Detroit Diesel 12V-149T engine, while the General Electric drive system powered electric traction wheel motors. Sadly for Dart, the DE930 and the 150-ton DE940 were commercial failures, mainly because their particular diesel-electric drivetrain set-up proved unreliable. The design of the DE930 as a working dump truck endured in later Dart models, even if the power system did not.

CHAIN-DRIVE TRANSMISSION

To an extent, the legendary Mack concern compromised its potential in the mining field by retaining its tried and tested chain-drive transmission system longer than most. It was undeniably a strong system but was considered archaic alongside Euclid's planetary gear transmissions. Mack soldiered on with its FCSW tandem-drive model – originally produced in 1937, and then right through to 1947 – and was thus overtaken by Euclid and its FD range of mining trucks. Mack saw the light in 1946 and its 15-ton LR series of 1946 and 22-ton LV series of 1948 were fitted with single planetary rear-drive axles. By 1949, it was offering the tandem-drive LRSW 30-tonner. Mack followed this up with its popular tandem-drive 34-ton LRVSW in late 1951, which carried it through the decade until it was superseded by the 40-ton LYSW in 1958. The following decade, the Mack M-series was the off-road model line, including the M-45SX, which appeared in 1960; the M-65SX, which followed in 1963; the M-70SX from 1965; and the M-75SX, which came out in 1970. There was also a trio of M-100SX trucks. The first was a 6x2 dump truck from 1966, powered by the 800bhp V12 Detroit Diesel 149T engine, with a carrying capacity of 100 tons. The second, made in 1967, was a tractor unit for pulling a

At launch in 1991, Caterpillar's 793 off-highway truck caused a sensation in the mining industry because it was the first 240-ton capacity rigid dumper that ran with a mechanical drivetrain rather than diesel-electric traction.

150-ton iron-ore trailer, and powered by a V12 700bhp Cummins VT12-700-Cl diesel, while the third vehicle used a 1000bhp Detroit Diesel and was bought by Hoffman Rigging and Crane Service of New Jersey in 1970 for use as a heavy haulage tractor.

Elsewhere, dump trucks were produced by Hino in Japan and Sisu in Finland, but the USA remained the habitat of the larger examples and home to the most manufacturers. International Harvester offered the 180 Payhauler, a 29-cubic metre (38-cubic yard) capacity dumper powered by a 475bhp V12 Detroit Diesel two-stroke motor with a 10-speed torque converter. But despite its prominence in commercial vehicle production, International Harvester couldn't match the performance of the specialists in off-road haulers.

While Euclid and Dart were the major players in the dump truck market during the 1950s, another key operator was getting geared up. Formed in 1953, LeTourneau-Westinghouse of Peoria, Illinois, was made up of the Westinghouse Air Brake Company (later WABCO) and the earth-moving equipment manufacturing part of R.G. LeTourneau's company. The company's prototype two-axle, rigid-frame, rear-dump hauler, the revolutionary LW-30 Haulpak, came off the production line late in 1956, and the model was available the following year. It had been designed by Ralph H. Kress, formerly executive vice-president and general manager of Dart Trucks, who many consider to be the founder of the modern dump truck. What was novel about it was the use of a Hydrair oil-air suspension system, also pioneered by Kress, and it had an offset cab layout and sloping, triangular-shaped dumper body. The Haulpak range of quarry and mining trucks went on to become the industry's benchmark design. Key features were the offset cab, Hydrair air-hydraulic suspension, and a triangulated dumper body that lowered the vehicle's centre of gravity. The LW-30 was powered by a 320bhp eight-cylinder Cummins diesel engine, with a payload capacity of 30 tons. This truck series became the Model 32 in 1959.

A lesser-known manufacturer was the Western Contracting Corporation of Sioux City, Iowa, which was moved to build its own 150-ton capacity Western 60 because it was unable to source anything large enough from elsewhere. This unique model was known as the Eucnik, and at the time of its introduction in 1958, it was the largest mining truck in the world. Load capacity of the Western 80 was a staggering 45.8 cubic metres (60 cubic yards), or up to 76.4 cubic metres (100 cubic yards) with heaped-up material. Western Contracting also operated Euclid 1LLD dump trucks in the 1950s, which they upgraded in 1960 by fitting them up with paired Detroit Diesel 12V-71 engines producing 650bhp.

DIESEL-ELECTRIC

Without exception, all heavy-duty, off-road mining trucks were equipped with mechanical drivelines consisting of an engine, manual or automatic transmission, and planetary gear drive axles. Then, in the 1950s, the diesel-electric powertrain began to appear. We're probably all familiar with the concept in terms of railway rolling stock, where it's been in common use since the disappearance of steam engines in the 1950s. But in dump trucks, this technology was largely promoted by R.G. LeTourneau once he had disposed of the earth-moving equipment portion of his company. In 1953, LeTorneau agreed to a five-year moratorium on producing machinery that would conflict with those of LeTourneau-Westinghouse, and in the interim, he developed his electric wheel traction motors specially for fitment in earth-moving equipment. The prototype electric-drive dump truck was the LeTourneau TR-60 Trolley Dump, first seen at his Longview, Texas, factory in 1959. LeTourneau developed the product in close cooperation with the mining research department of the Anaconda Corporation, and the vehicle became operational at the company's Berkeley Pit in Butte, Montana, in 1960. The original articulated TR-60 was powered by a 335bhp Cummins Super-Diesel engine, to which AC and DC generators were hooked up. In turn, these provided the power for four electric traction motors which were located in each wheel hub assembly. But when the TR-60 was connected to overhead electric wires via its front-mounted pantograph arms, the electric wheel motors could produce the equivalent of 1242bhp. The purpose of the Cummins engine was to make the vehicle independent of the trolley cables, which didn't necessarily extend to wherever the dumping ground was. The Cummins had to be running constantly while the vehicle was in operation, because it supplied power to the braking and steering system, even when it was hooked up to the overhead cables. The 60-ton capacity TR-60 was later fitted with an additional 335bhp diesel engine in order to power the generator, necessary to augment the cable power when operating at low speed and when heavily laden. Payload capacity was thus boosted to 75 tons in the process.

By the mid-1990s, Euclid was building an uprated version of its R-190, known as the R-220 dump truck, which also featured the pantograph arrangement mounted on a platform constructed ahead of the front of the vehicle. The regular powerplant was the Cummins K2000E or optional Detroit Diesel 16V-149TIB, and carrying capacity was 217 tons. A fleet of R-220s was operational at the Palabora Copper Mine in South Africa from 1995.

One company that would become a major manufacturer of earth-moving plant was the Unit Rig & Equipment Company of Tulsa, Oklahoma. It had been a manufacturer of oil-drilling equipment since 1935, and in 1960 it constructed a diesel-electric dump truck for use in open-cast mining sites. Previously its extensive experience with powerful diesel engines and electrically driven machinery systems was limited to drilling rigs. The prototype was known as the Lectra Haul M-64 Ore Hauler. This designation rolls off the tongue quite nicely, but its meaning is a bit garbled. What it meant in practice was that the vehicle used the General Electric motorised wheel system, and was a 64-ton capacity, articulated-steered, rear-dump hauler. Notably, it was the first dump truck to use the General Electric system and was otherwise powered by a 630bhp Cummins VT-12-BI diesel engine which powered the main generator. This in turn drove the electric motors located within the wheel rims. Tests at the Hanna Mining Company's iron ore mine in Minnesota led Unit Rig to conclude that the vehicle's articulated design wasn't suitable for operating in open-cast mines, despite the fact that the diesel-electric driveline worked very well. Results of the test-bed M54 filtered through to Unit Rig's next dump truck, the Lectra

The engine and chassis of a Caterpillar 777 dumper. These are massive bits of machinery. Count the cylinders on the engine. It is a 'V' so you have the same number at the other side. Note also the ladder attached to the engine.

DUMP TRUCKS

O&K K60

Make: *Orenstein & Koppel AG*
Model: *K60 rigid dump truck*
Manufactured: *1997*
Engine: *Cummins KTA 19-C 675*
Transmission: *Allison 6f/2r*
Power: *504bhp*
Capacity: *60 tons*

Haul M-85. The original articulated steering chassis was superseded by a rigid frame, with diesel-electric drive that incorporated General Electric traction motors in the rear-drive wheels. It had an 85-ton payload rating and power was supplied by a 700bhp Cummins VT-12-700 engine. In its day, it was one of the biggest two-axle dump trucks on the market, and it can be regarded as the first diesel-electric rigid-frame rear-dump truck to be

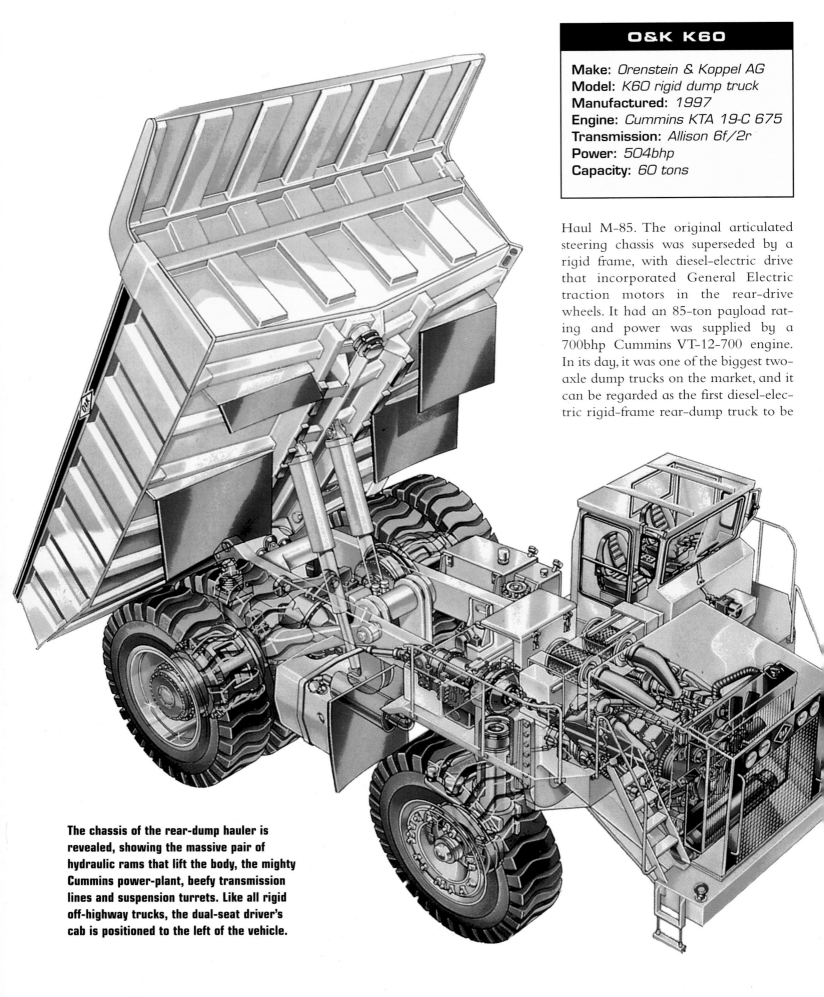

The chassis of the rear-dump hauler is revealed, showing the massive pair of hydraulic rams that lift the body, the mighty Cummins power-plant, beefy transmission lines and suspension turrets. Like all rigid off-highway trucks, the dual-seat driver's cab is positioned to the left of the vehicle.

produced specifically for mining applications. The first M-85 was sold to Kennecott Copper Co in 1963 and went to work at the Chino mine in New Mexico. At a stroke, the reputation of Unit Rig's Lectra Haul dump trucks was established in the mining equipment industry.

SIZE MATTERS

In 1967, Unit Rig's Lectra Haul M-200 model was the first 200-ton capacity diesel-electric drive production truck to go into service in the mining industry. Unit Rig went on to produce 120 units of this model, which gives an idea of the scarcity of dump trucks of this size and stature. The M-200 dump truck was introduced in 1966, with initial production fulfilling an order from

Articulated dump trucks like this Scania-powered six-wheel drive Moxy MT36 have the advantage of greater versatility than rigids as they can access tighter turns and have greater axle articulation on rough terrain. Capacity is generally much less than rigid dumpers, however.

Kaiser Resources of Sparwood, British Columbia, to be put to use in its opencast coalmines. At first, they were powered by a 1500bhp eight-cylinder General Motors EMD 12-645-E4 diesel locomotive engine, progressing to a 2250bhp 12-cylinder GM EMD 12-645-E4 unit. Initially the vehicles suffered from adverse tyre wear, but larger 40.00-57 x 6OPR boots provided the cure. The Lectra Haul M-200 might have been Unit Rig's biggest dump truck during the early 1970s, but its most successful model was the Lectra Haul Mark 36, which had a more modest 170-ton capacity. Whereas they only made 120 M-200s, there were a total of 685 M-36s in use at numerous mine and quarry sites worldwide. Main power units were the Cummins KTTA50-C1600, or the 16-cylinder Detroit Diesel 16V149TI, both providing diesel-electric power rated at 1454bhp. In the following decade, Unit Rig produced successively bigger dump trucks, including the MT-1900 of 1984, the MT-2050 in 1986, and the MT-2120 in 1987. By 1998, the company was active in what had become the more competitive 240-ton class with its Lectra Haul MT-4000. This machine could be specified with a choice of three engines, the largest of which was the 2334bhp 20-cylinder Detroit Diesel 20V-149TIB. Specially equipped versions ran with dual radiators with parallel cooling fans, necessary if the vehicles were working at sites in hot countries. The MT-4000 measured 13.1m (43ft) long and 7m (23ft) wide, weighed 140,280kg (309,266lb) empty, and had a ceiling of 358,005kg (789,268lb). It was shod with 40.00-57,66PR tyres. Larger still was the Unit Rig Lectra Haul MT-4400 that came out in 1994. First units off the production line went to the Powder River Coal Company's Caballo coalmines in Gillette, Wyoming. The MT-4400 should have been rated as a 280-ton dump truck, but when its intended 1.1-m (44-in) width tyres never appeared, they were down-scaled to 260 tons capacity. It was powered by a 2267bhp V16 Detroit Diesel/MTU 16V-396TE engine, and had an empty weight of 156,295kg

DUMP TRUCKS

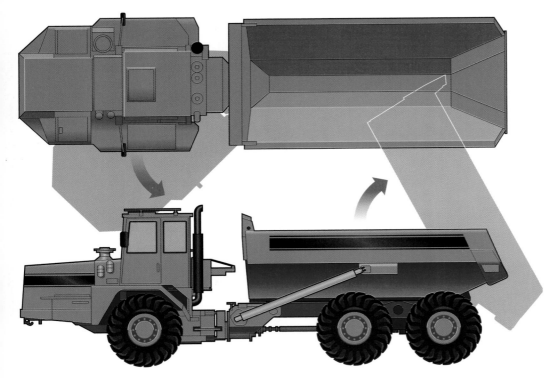

Side and aerial view of an articulated dumper. The single-axle tractor format makes for a very tight turning circle, while the hopper, almost an inverted triangle, is elevated by a pair of long hydraulic rams on either side.

(344,572lb) or 392,357kg (865,000lb) fully laden. The MT-440 measured 13.89m (45ft 7in) by 7.36m (24ft 2in).

ARTICULATED DUMP TRUCKS

Alongside the rigid haulers is their close relative, the articulated dump truck. Broadly speaking, it's a development of the tractor and trailer concept. Take a rigid model and add an articulation joint and an oscillation ring between cab and dump body, and there you have an articulated dump truck. The articulation joint enhances mobility as it allows the vehicle to bend in the middle, and the oscillating ring permits the cab to move independently, allowing a good range of axle articulation when travelling across rough terrain. This makes them rather more versatile than rigids, although less robust. The hydraulic ram and hoses tend to be more exposed, and therefore more vulnerable to damage. The typical articulated profile has the engine located far forwards in the frame, with the cab positioned directly over the front axle. Among the first such dumpers was the Northfield Engineering 4x2, unveiled in the UK in 1961, which had a 12-ton payload. By the 1990s, the manufacturer of the largest volume of articulated dump trucks was US giant Caterpillar – of whom more later – who produced a range of six all-wheel drive models. Four of these were 6x6 and the other two were 4x4s. Biggest was the D400E, powered by the 298bhp Cat 3406 engine and featuring hydraulic front suspension and a Hydroflex rear set-up with hydraulic damping and load-transfer control. A measure of its versatility was its compact 3.96m (13ft) turning circle, while the 25-ton maximum capacity payload shows up the limitations of the articulated format. The Scania-powered MT 40 articulated hauler, produced by the unlikely alliance of the Japanese-US-Norwegian concern Komatsu Moxy, had a capacity of 44 tons, while the Terex 4066C that used a Detroit Diesel motor was also good for 44 tonnes. By the late 1990s, Volvo had the distinction of being the most prolific builder of articulated dump trucks. At the lower payload capacity end of the range was the 6x6 Volvo A25C, powered by the 195bhp Volvo TD73 KCE engine.

The most ubiquitous of all heavy plant makers was Caterpillar, which alone amongst dump truck producers always made its own componentry. With origins as far back as 1910 in the Holt steam crawler tractors, Caterpillar made its first off-road truck in 1962. It was designated the 769, and was a 35-ton rear-dump hauler of 35-ton capacity. It had mechanical transmission and a power-shift. By now, Caterpillar had hired the services of Ralph Kress, formerly of LeTourneau-Westinghouse, as its new Development Manager, and his electric-drive system was installed in the prototype for the upcoming 75-ton capacity Model 779. It also featured in the Caterpillar side-dump three-axle Model 783 of 1965, as well as the enormous Model 786, which was a 240-ton bottom-dump coal hauler. The common power unit was the 960bhp V12 Caterpillar D348 engine, linked to in-house electric-drive systems, and there was much interchangeability of parts, on account of the concurrent design programme for all three models. However, the electric-drive system wasn't as reliable as it might have been, and there was a move towards redirecting technical resources towards the development of a power-shift automatic transmission system for fitment in a 50-ton hauler. When Kress retired in 1969, Caterpillar's diesel-electric programme was cancelled, and all its subsequent dump trucks ran with conventional mechanical-drive.

BOTTOM-DUMP COAL HAULERS

This esoterically named category of dump truck does just what the name suggests: it off-loads from its underside. It is found almost exclusively in the US and Australian coalmines, and the manufacturers who service this sector also make rear-dump trucks. There are a few other dedicated specialists in the field; thus we find Euclid, Terex, Caterpillar, Komatsu, Dart and Unit Rig, as well as Rimpull and Kress. As the vehicle type evolved, certain models were more suitable for working with

DUMP TRUCKS

soil – known as overburden – while others carried coal or salt.

Like the articulated versions of the rear-dump haulers, bottom dump vehicles are used for carrying lighter weight overburden material, but their particular configuration, like the bomb doors on a bomber aircraft, means they can handle a larger volume than the rear-dump haulers. In many cases, the bottom-dump's tractor unit is derived precisely from that of the rear-dump hauler of the same family.

IT'S A GAS

Around the mid-1960s, experiments were made with gas turbine units in the quest for more powerful engines. Gas turbine engines had the advantage of a high power output and lower engine weight than diesel units, plus good reliability and long intervals between servicing. The concept was tried by car manufacturers Rover and Howmet on the race-and-test track with spectacular, if unenduring, effect. However, the major deficit was that they used probably twice the fuel of a regular diesel unit. The instigator of gas turbine trucks was Unit Rig who came out with two modified Lectra Haul M-100 dump trucks in 1965. One was powered by a 1100bhp General Electric LM-100 gas turbine, while the other used a 1000bhp IH Solar Saturn gas turbine. In 1969, five 120A Haulpak dump trucks powered by IH Solar Saturn gas turbines were evaluated by WABCO, the new corporate identity for LeTourneau-Westinghouse. While test results were promising, they were beset with numerous problems to do with the adaptation of the powerplants to existing truck designs. As it turned out, the majority were restored to their correct diesel-electric format.

Under the auspices of its parent company, the White Motor Corporation, Euclid installed a 1000bhp Avco-Lycoming TFI4 gas turbine motor in a converted R-105 articulated dump truck in 1970 and went to play in the Mesabi Iron Range of Minnesota and at its Laredo, Texas, proving ground. Although the vehicle was given back its original powertrain, the data accumulated on tests was used when it came out with the Euclid R-210 Turbine Hauler in the early 1970s. In fact this was the only mining dump truck designed to run solely on turbine-electric power, and it had the distinct advantage of a very good power-to-weight ratio thanks to the engine weighing less than half a regular diesel engine. The R-210 prototype used the compact Avco-Lycoming TF25 engine that pushed out no less than 1750bhp to a quartet of electric traction motors mounted in each wheel. Load capacity was 210 tons, while unladen weight was 111,130kg (245,000lb) or 301,638kg (665,000lb) fully laden. Unusually, the R-210's engine bay was between the chassis frame rails, mounted behind the driver's deck and just in front of the rear-axle housing. The gas turbine powered the generator, which

Introduced in 1996, Rimpull's magnificent CW-180 bottom-dump coal-hauler is powered by an 1100bhp V12 Detroit Diesel engine. It can carry 180 tons of coal and weighs 300 tons fully laden. Rimpull's biggest model, the CW-280 can accommodate 280 tons and is a truly immense vehicle.

DUMP TRUCKS

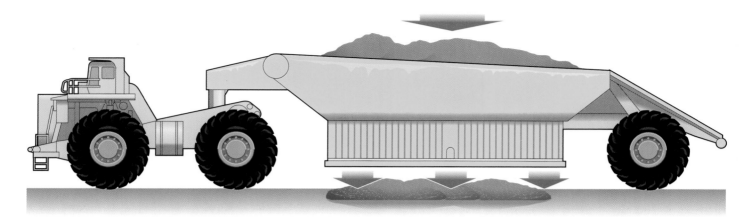

in turn drove each of the traction motors in the wheels. The R-210 debuted at the American Mining Congress show in Las Vegas in October 1971, and although tests at the Laredo proving-grounds during the following two years went well, it was effectively sunk by the oil crisis of 1974. Fuel prices rocketed, and of course a guzzler like the R-210 stood no chance. The prototype was sold to the Bougainville Copper Mine in Papua New Guinea in 1974, and that was that for gas turbines.

DIESEL-ELECTRIC POWER

When it was clear that gas turbine power was not the way forward, heavy-plant manufacturers turned to the only other source of powerful engines, which was to be found in the diesel locomotive. Such engines produced massive amounts of power at very low revs, and although they were very heavy – which accounted in part for their robustness – they were reliable. The first production dump truck with a locomotive engine installed as standard equipment was Unit Rig's Lectra Haul M200, introduced in late 1968.

Another attempt at building a viable diesel-electric dump truck was made by Dart, who came out with the prototype DE-2291 in 1970. Its weighty 1800bhp eight-cylinder General Electric FDL-8 locomotive motor proved to be just too heavy for the front-wheel steering system, causing scrubbing damage to the tyres while the vehicle was on performance trials at Kaiser Steel's Californian Eagle Mountain Mine.

In 1970, the biggest dump truck in the mining industry was the otherworldly Peerless VCON 3006. It was a 250-ton capacity vehicle with four front and four rear wheels, side radiators and a load bed activated by massive hydraulic rams, and looked like something off the set of Starwars. Designed and built by the Peerless Manufacturing Company of Dallas, this colossal prototype was dispatched to the Pima Copper Mine in Arizona for tests in June 1971. The Peerless Manufacturing Company's normal field of operations was to make specialised equipment for the natural gas transmission and petrochemical industry, and it had founded a Vehicle Construction division in 1970. The VCON 3006 was powered by a 2790bhp 12-cylinder Alco 251-12E diesel locomotive engine, which supplied six General Electric traction motors. These were located one in each of the four rear wheels, with one motor in each of the two outside front wheels. A measure of the breadth of the scope of some manufacturers is that Alco, for instance, also made cement mixers. The Alco-powered VCON 3006 designation stood for 3000hp and six driving wheels, which my backyard mixer most definitely does not have. The VCON 3006 was based on a walking beam chassis frame with hydrolastic suspension and independent hubs. It was 13.1m (43ft) long and 8.5m (28ft) wide, and weighed 158,757kg (350,000lb) unladen, and 385,553kg (850,000lb) fully laden.

By May 1973, the Marion Power Shovel Company had bought the

The bottom-dump coal-hauler is a gigantic articulated dump truck, and works by opening the doors in the underside of its hopper to release the material – usually coal, but overburden and in special circumstances, salt is carried. Some units produced by Rimpull and MEGA operate with tandem trailer units for greater capacity, while rigid-chassis versions are made by Kress Corporation.

VCON division of Peerless, and with interests in other areas of earth-moving equipment, the funds for developing the VCON 3006 were diverted to further the VCON V-250 Dozer scheme. When the Russian Coal Ministry expressed an interest in the VCON 3006 in 1970, Marion had proposed a joint venture with Sumitomo Industries to produce the vehicle. However, the Russian Coal Board selected the Lectra Haul M-200 instead, although the package was to have included a number of Marion 204-M Superfront Shovels, built by Sumitomo Industries. Corporate politics don't fit too well with heavy plant machinery, and when Dresser Industries acquired the Marion Power Shovel Company in 1977, all development work on the mighty VCON 3006 dump truck came to a halt.

Another giant of a machine was the 200-ton capacity WABCO Haulpak 200B, which in 1969 was the largest diesel-electric driven three-axle articulated rear-dump truck that had ever been available. It was powered by the 1440bhp 16-cylinder General Motors 16V-149T diesel engine, which powered the rear axle of the tractor unit

DUMP TRUCKS

and the front axle of the triangulated dumper trailer, effectively making it a four-wheel drive vehicle, although there was an unpowered axle. The 200B had a maximum laden weight of 306,174kg (675,000lb), and it was clearly of a different configuration so was longer than average at 16.7m (55ft) and of a narrower 6.1m (20ft) width.

HAULPAK LINE

In 1971, WABCO began testing the prototype for a run of six 200-ton Haulpak 3200 diesel-electric tandem-drive three-axle dump trucks. The chosen power unit was the two-stroke, 1800bhp 12-cylinder, GM EMO 12-645-E4 diesel locomotive engine. Maximum all-up weight of the 3200 was 325,225kg (717,000lb), and being a long-wheelbase three-axle vehicle, length was 15.3m (50ft 6in) and width 6.8m (22ft 7in). An upgraded 2250bhp version was introduced in 1975, with carrying capacity upped to 235 tons, partially by fitting larger 36.00-51 x 42PR tyres. This was the WABCO 3200B, and during the second half of the 1970s, it was the largest-capacity version of the Haulpak series to be built. Its maximum payload was increased to 260 tons with an unladen weight of 165,561kg (365,000lb) or 401,429kg (885,000lb) fully laden. It was slightly longer and wider than its predecessor, at 16.4m (54ft) in length, and 7.31m (24ft) wide. A total of 48 units were built before the 3200B went out of production in 1983.

All aspects of the automotive and commercial vehicle industry are subject to shifting fortunes and corporate merger and takeover. The heavy plant sector is no exception. Thus we find Komatsu building the Haulpak range in the late 1980s. The 830E model dump truck was launched in April 1988 and designated the Dresser Haulpak 830E. It quickly became the best-selling diesel-electric truck in the 240-ton capacity class, and was still in production in the late 1990s. And, wonder of wonders, it was available in off-white, rather than the ubiquitous orange-yellow. It was 13.5m (44ft 4in) long and 7.31m (24ft) wide, and maximum payload capacity was increased to 255 tons. The standard power unit was the 2409bhp 16-cylinder MTU/DDC 16V-4000 engine. The 830E ran on 40.00-57-size tyres, and unladen weight was 154,401kg (340,398lb), or 385,848kg (850,650lb) fully laden.

Coming much more up to date with Komatsu's 310-ton capacity Haulpak 930E, launched in May 1995, we find that it runs with an AC electric-drive motor system. Previously, all diesel-electric dump trucks used DC traction motors. The efficient General Electric AC drive system embraced many of the advanced control systems and components that GE had developed for their AC diesel-electric railway engines. At any rate, the 930E's main power source was the 2500bhp V16 MTU/DDC 16V-4000 diesel

Unit Rig launched the 260-ton MT-4400 mining truck in 1994. It's powered by a 2287bhp Detroit Diesel/MTU 16V-396TE engine. The giant air cleaners give the impression of headlights, but those are set low down in the radiator grille.

Cutaway of the Terex 4066C highlights the forward location of the Detroit Diesel engine, relationship of tractor unit to articulated all-wheel drive chassis, and double axle supporting the hopper unit.

TEREX 4066C

Make: *Terex*
Model: *4066C ADT*
Manufactured: *1997*
Engine: *6-cylinder Detroit Diesel Series 60*
Power output: *298bhp*
Carrying capacity: *21.5 cubic metres (28 cubic yards)*

engine, which was wired up to the AC motors.

TEREX TRUCKS

One of the most important names in the field of earth-moving vehicles was Terex. This was a division of the General Motors Corporation, and Terex launched its first diesel-electric dump truck, the 33-15 series, in May 1971. The Terex 33-15 was powered by a 1445bhp V16 Detroit Diesel 16V-149TI engine, and weighed 242,671kg (535,000lb) fully laden. It was built at the company's EMD plant at London, Ontario, and the first 150-ton capacity vehicle was uprated to 170 tons in 1975 as the 33-15B, which lasted until 1981. That, coincidentally, was when General Motors sold off Terex to IBH, although confusingly, GM retained ownership of the Canadian plant and changed the name of its product to Diesel Division Titan trucks. Terex, meanwhile, operated from plants at Motherwell near Glasgow, Scotland, and Tulsa, Oklahoma. It had a larger 200-ton capacity two-axle dump truck ready for launch in 1980, but the project was stymied due to prevailing economic circumstances. Resplendent

DUMP TRUCKS

in a lime-green hue, Terex went its own way with a series of mining trucks including the 3345, 3360 and 33100 Off-Highway haulers. The latter was powered by a 783bhp Cummins KTA 38-C engine, measured 11.1m (36ft 6in) long, and had a payload capacity of 100 tons. Like the majority of these vehicles, it had a box-section high-tensile steel chassis, with added castings reinforcing high-stress areas of the frame. The front suspension struts were fixed to a closed loop frame member, and the dump body was hinged to the rear of the frame: pretty basic engineering, but very effective.

The Terex 33-19 Titan was the biggest dump truck anyone had ever seen at its launch in 1974 at the American Mining Congress show in Las Vegas. With a staggering 350-ton capacity, the three-axle 33-19 Titan was powered by a 16-cylinder two-stroke GM EMD 16-645-E4 diesel locomotive engine of no less than a 169-cubic litres (10,313-cubic inches) capacity. That's a colossal engine, and since all two-stroke diesels are supercharged, it would have sounded fantastic, and yet it delivered its 3000bhp at a truly leisurely 900rpm. Unsurprisingly, this was the largest locomotive-type engine application ever fitted in a dump truck, and it was to be the last. The 33-19 didn't use wheel motors, and instead, its four electric GM D79CFA traction motors were mounted inside the rear-drive axles, with two in each unit. In fact, the Titan was configured around the EMD engine, and its 10 wheels were shod with the largest tyre size available: 40.00 x 57,60PR. The Titan went to work at Kaiser Steel's Eagle Mountain Mine in southern California.

The Terex 33-19 Titan of 1978 was a massive vehicle in every dimension. It was 20.1m (66ft) long, by 7.79m (25ft 7in) wide, and weighed 235,868kg (520,000lb) unladen or 553,564kg (1,220,400lb) full up. Perhaps surprisingly, it could just about make 48.2km/h (30miles/h) on the flat with a full hopper, which seems quite fast considering its bulk. It was equipped with rear-wheel steer characteristics, which assisted manoeuvrability and helped redress a tendency for big three-axle dump trucks towards understeer. The model was first operational in late 1978 in the Rockies, in Kaiser Resources' Balmer Mine at Sparwood in British Columbia.

After General Motors disposed of its Terex Division in 1981, the Terex nomenclature ceased to feature on the 33-19 and it was replaced with the Titan name, which came to represent the whole of GM's Canadian Diesel Division. The vehicle itself was repainted in the colours of Westar Mining in 1984 after the Kaiser mine where it worked was taken over by West Mining in 1980. The Titan 33-19 was operational until 1990, when it was sidelined by a badly cracked rear axle housing. Its fate was not too ignominious, as it ended its days as a permanent exhibit in Sparwood, courtesy of the mine owners. Although there were provisional orders for further examples of the 33-19 dump truck on the books in 1981, they never amounted to anything largely because of the worldwide recession that was brewing. Thus the 33-19 was a one-off.

At the beginning of 1985, General Motors' Canadian Diesel Division sold off the Titan marque to the Longview, Texas-based Marathon LeTourneau company. At first, the product was billed as the LeTourneau Titan 33-15C and was not really any different from the General Motors-built dump truck. However, shortly after making the acquisition, Le Tourneau set about revamping the model line, which ranged in capacity from 170 to 240 tons. This excercise culminated a decade later in its T-2000 series of TITAN dump trucks. Note the use of capital letters now. With a haulage capacity of 200 tons, the T-2200 was powered by 2000bhp Cummins or Detroit Diesel engines, and weighed 143,743kg (316,900lb) void or 325,180kg (716,900lb) fully laden. LeTourneau's largest capacity dump truck was the Titan T-224D, capable of transporting 240 tons of matter. The first two prototypes were dispatched to the Westar Mining Company in British Columbia, Canada, for evaluation in 1990. They were powered by the 220bhp 16-cylinder Detroit Diesel 16V-149TI engine, and weighed 152,815kg (336,900lb) unladen or 370,539kg (816,900lb) filled up. It was

The South African company Bell Equipment make a range of articulated rear dumpers, including this B20-C 6X4, tipping mineral-bearing rock into a crusher at Bridgestone Dolomite mine at Moorreesburg in Western Cape province.

14.3m (47ft) long, 6.7m (22ft) wide, and was shod with 40.00-57,68PR tyres. Production lasted until 1995; just four units of this model had been made.

The enormous tyres vital to the functioning of these big trucks are almost as specialised as the vehicles that wear them. They might look big and crude, with none of the complex tread patterns that you get with high-speed tyres, but understandably, they have to be robust and rigid enough to withstand the hammering they get in a quarry or mine, yet supple enough to absorb humps and undulations. In this way they also contribute to the dump truck's ride quality. One such vehicle that was fitted with the largest off-road radial tyres was the Komatsu Haulpak 930E. The tyres in question were Bridgestone 48-95R57 VELS. But premature wear-rates led the tyre manufacturers to think again, and Bridgestone returned with its broad-shouldered 50-90R57 series radials. In the case of the Haulpak 930E, the wider tyres enabled the dump truck's carrying capacity to be raised to 320 tons. This meant that, fully loaded, the truck weighed 480,354kg (1,059,000lb), and the practice of fitting wider tyres often meant that the carrying capacity was lifted.

The vast, remote regions of the former Soviet Union were an ideal proving ground for mining haulers, but like the Kamaz trucks produced there, heavy plant machinery was rarely seen in the West, at least not until the collapse of the political regime. One such manufacturer was the Minsk-based Byelorussian Automobil, which made BelAZ dump trucks at its Zhodino auto works plant in the Republic of Belarus. Its biggest offering was the 308-ton capacity BalAZ 75501 ultra-hauler, begun in 1988 and introduced in 1991. Its configuration was somewhat at odds with conventional dump trucks. It was a two-axle vehicle, with two pairs of wheels on each axle, and articulated at the centre, by which means it was steered. The BelAZ 75501 was powered by a Russian diesel-electric locomotive engine designated the 12ChN1A 26/26, which was wired up to the DC electric traction motors. Originally, a small amount of the hauler's componentry was sourced from Komatsu, and this extended to the operator's cab, the tipper's hydraulics, and diagnostic systems, but this source was threatened by the conflict of interests surrounding Komatsu's tenure of the Haulpak range of dump trucks. The prototype BelAZ 75501 was last heard of operating in an open-cast coalmine in Nerungri in the depths of Siberia.

In the late 1990s the UK's main producer of rigid dump trucks was Aveling Barford, a Lincolnshire firm that had a long history in the manufacture of construction industry vehicles. Rather smaller in scale than their US counterparts, Aveling Barford's largest offering was the 65-ton capacity RD65, while its smallest was the 30-ton RD30. The RD65 was powered by a turbocharged

DUMP TRUCKS

aftercooled Cummins straight-six diesel engine linked to an Allison ATEC6062 transmission with TC683 torque converter. Aveling Barford also made a couple of articulated haulers designated the RXD 24 and RXD 28, both powered by the 205bhp Cummins LT 10 diesel. Yorkshire-based DDT Engineering was a small volume maker of articulated six- and four-wheeled articulated dump trucks of 29 tons payload that used Cummins power and a moving headboard discharge system. Another British maker was Heathfield Haulamatic, whose H44 and H50 rigid rear-dump trucks used proprietory Cummins KTA 19C engines with Allison CLBT 5963 ATEC transmissions. Payload capacities were, predictably, 44 and 50 tons respectively.

In the same league were the rear-dump haulers from Germany's long-standing engineering concern Orenstein and Koppel AG (O&K for short.) Like the British manufacturers mentioned above, O&K used proprietary US-made Cummins engines and Allison transmissions in its K45 and K60 dump trucks. Its largest product was the 11.12m (36ft 6in) long, 100-ton capacity O&K K100, powered by the 783bhp Cummins KTA 38-C 1050 unit, and shod with 27-R49 XHDIA tyres.

By 1998, Terex had absorbed Unit Rig & Equipment, and expanded its line of rear-dump haulers to six models, including the medium-duty TR45. Maximum payload was 45 tons.

MAJOR THRUST

Evolutions in the world of vehicle production are quite often cyclic, and the major thrust in the early 1980s was towards productionising a two-axle dump truck with a carrying capacity of 200 tons, formerly the province of the long wheelbase three-axle models. The first dump truck to exceed the 200-ton capacity threshold was the Wiseda KL-2450, otherwise known as King of the Lode. One William Seldon Davis, a scion of the family that owned and operated Unit Rig and built Lectra Haul vehicles, launched the 220-ton capacity King of the Lode in 1982, having recently set up Wiseda to manufacture high-capacity dump trucks. Fans of the bizarre acronym will notice that the company name was composed of elements of William Davis's own names. In 1985, Wiseda followed up its ground-breaking advance with the first 240-ton capacity mining dump truck, the KL-2450. A decade later the vehicle was being produced by Liebherr, who changed the designation to the T-282 in 1997. Among the seven-engine options available was the 2250bhp MTU 16V-396 TE44 diesel unit. Laden weight of the Leibherr T-282 was 369,759kg (815,180lb), with higher capacity vehicles made to special order.

The industry giant, Caterpillar, was a little slower off the mark in heavy-duty mining trucks, preferring to feel its way with well-proven designs. Its Model 785 of 1984 was a 150-ton capacity vehicle, and this was followed two years later by the 195-ton Model 789. In the late 1990s, this Model series was still dominant in the 190- to 200-ton sector.

TEREX TR45

Make: *Terex*
Model: *TR45 off-highway rear-dump truck*
Manufactured: *1998*
Engine: *Cummins KTA 19-C*
Power: *363bhp*
Transmission: *Allison ATEC 6f/2r*
Capacity: *45 tons*

Meanwhile, the products on offer throughout the mining industry continued to increase in stature and composition. Among the bigger ones were the Komatsu 730E at 205 tons, and the similar capacity Unit Rig MT-3700B. Next up were the 215-ton Liebherr T-252, the slightly larger 217-ton Euclid R-220, and bigger still, the 240-ton Caterpillar 793C and 250-ton Liebherr T-262 and 255-ton Komatsu 830E. The Unit Rig MT-4400 could haul 260 tons and the Euclid R-260 managed 262 tons. The one that most operators seemed to go for though was the Caterpillar 793/B/C series, which came out in 1990, and had a mechanical powertrain, rather than the electric traction type, indicating the operators' preference for a traditional driveline. In fact, up to this point, no other manufacturer had built a dump truck of this size with a mechanical transmission and no electric wheel motors. By 1993, an uprated version – the Caterpillar 793B – was up and running, and within just three years the company had sold over 550 units all over the world. The 793C used Caterpillar's own 2057bhp 3516 V16 diesel engine, which drove through an electronically controlled six-speed automatic Cat transmission with power shift. Total laden weight of the 793B was 376,481kg (830,000lb). The Caterpillar 793C which came out in mid-1996 utilised the leaner, cleaner 2166bhp V16 Cat 3516B EUI V-16 diesel unit, and the vehicle was rated at a 240-ton payload capacity.

Powered by the 1705bhp Caterpillar 3516 V-16 diesel engine, the Caterpillar 789 series remained the world leader in the 195-ton payload class. This too used a transmission and drive shaft driveline instead of a generator and electric traction motors, and its successor was the upgraded Model 789/B launched in 1992, followed by the 789C in October 1998. All-up weight of the 789 was 317,514kg (700,000lb).

ULTRA HAULERS

A new race of mining dump truck began to appear in the mid-1990s. These vehicles were considerably bigger than their predecessors, with payloads around the 300-ton capacity rating. Vehicles with this kind of capacity had previously been the triple-axle variety, and the new breed of two-axle six wheeler came to be identified colloquially as the ultra hauler. The days of the lumpen, slow-revving diesel-electric locomotive engines were over, and the incoming brand of ultra-hauler dump trucks were powered by high-revving diesel engines and featured the very latest in tyre technology. A fine example of this was the Komatsu Haulpak 930E, announced in 1995, which was an advance on previous incarnations of the Haulpak line. As we've seen, it evolved during a succession of corporate name changes, mergers, joint ventures and acquisitions under the aegis of a number of companies including LeTourneau-Westinghouse, WABCO, Dresser, Komatsu-Dresser, and finally, Komatsu. By the end of the century, the largest Haulpak dump truck was the 930E, rated at first as a 310-ton capacity vehicle, and soon uprated to 320 tons. The main innovative feature of this vehicle was its AC traction-drive system, which had been developed jointly by Komatsu and General Electric. Although it was widely used in railway engines, this was the first use in a dump truck, and by the year 2000, around 100 units had been built. Another dump truck to use an AC-current diesel-electric drive package was Liebherr Mining Equipment's 360-ton T-282. The electric part of the drive system was made by Siemens Industries. Liebherr entered the dump truck sector in 1995 after buying Wiseda from William Davis. The KL-2640 was initially rated at 320-tons capacity, evolving in 1997 into the 340-ton KL-2680, and the following year the designation became the T-282. Unveiled at the company's Baxter Springs plant in 1998, it was fitted with the latest 1.6m (63in) wheel rims and shod with 55/80R63 tyres, rather than the 1.4m (57in) rims that were previously the industry norm. The height of a wheel and tyre was now about 3.65m (12ft), and the width 1.21m (4ft), which reflected the need for bigger boots to support ever-increasing load capacities.

Further developments in hand at Liebherr Mining Equipment at the turn of the new millenium included the TI-272, otherwise known as an Innovative Large Mining Truck – or ILMT – which was evolved in collaboration with the Australian mining company Broken Hill Properties. The project was commissioned by the Central Queensland Coal Association in 1996 for the Saraji Coal Mine in

DUMP TRUCKS

Queensland, Australia, which is operated by Broken Hill Properties Coal Pty. Originally designated the IL-2600, the TI-272 dump truck's evaluation trials began in 1996, carrying a 240-ton payload, but a capacity for the production vehicles was targeted at 300 tons. The diesel-electric prototype was powered by a 2250bhp MTU 16V-396 TE44 engine, with a fully laden weight of 351,534kg (765,000lb). At the root of the experiment was the replacement of the differential housing in the rear axle that normally would have joined the

Liebherr's 360-ton T-282 ultra hauler dwarfs the pickup at the Powder River Basin Black Thunder mine, Wyoming. Introduced in 1998, the T-282 is powered by a 2750bhp Detroit Diesel MTU/DDC 16V-4000 turbo-diesel engine, driving Siemens/Liebherr AC electric traction wheel motors.

LIEBHERR T-282

Make: *Liebherr Mining Equipment*
Model: *T-282 ultra hauler*
Manufactured: *1998*
Engine: *Detroit Diesel MTU 16V-4000 turbodiesel*
Power output: *2750bhp*
Carrying capacity: *360 tons*

rear frame structure together. Instead, the dump body acted as an integral part of the frame structure, and with electric traction motors in place, the rear wheels could now be spaced equally across the truck's back end, with suspension struts mounted on individual axle boxes in the manner of the Peerless VCON 3006 prototype. The hod – or dump box – became part of the chassis structure, and the weight saved was considerable, estimated in fiscal terms to be worth approximately 15 per cent of the cost of the truck's lifespan.

RECORD BREAKER

Caterpillar's mammoth 797 dump truck had a record-beaking carrying capacity of 360 tons, and like its lower-rated siblings, was equipped with a

One of the most popular rigid rear-dump haulers is the 240-ton capacity Caterpillar 793C, announced in 1996, and running with a V16 Cat 3516B diesel engine rated at 2166bhp and mechanical transmission. The cab is accessed via the staircase running up the radiator grille.

mechanical powertrain, based on a 3250bhp 24-cylinder Cat Diesel 3524B LS EUI motor. Utilising some of the principals of modular engine systems, this power unit was created by linking a pair of Cat 3512B engines at the crankshaft via a flexible coupling system. Transmission to the rear-mounted differential was through a computerised seven-speed automatic gearbox. Just as the Liebherr T-282 was designed to run on tall 160cm (63in) rims, these were also part of the specification of the Caterpillar 797, which was shod with 55/80R63 tyres. It weighed 557,918kg (1,230,000lb) fully laden, and measured 14.5m (47ft 7in) by 19.4m (30ft) wide. No other two-axle dump truck had ever been as wide. In late 1998, the prototype 797s were put through their shakedown tests at Caterpillar's proving grounds in Arizona.

By 1997, Unit Rig had become part of Terex Mining, and work began on the prototype of its Lectra Haul MT-5500. This ultra-haul dump truck was powered by a Siemens electric traction motor system with AC current, developed by Unit Rig in consort with the Power Conversion Systems Division of General Atomics. Like the Caterpillar and Liebherr models, the Lectra Haul MT-5500 ran on the same 160cm (63in) wheel rims and 55/80R63 tyres, and its carrying capacity was rated at 340 tons.

Now joined with the Japanese Hitachi concern as Euclid-Hitachi Heavy Equipment, Euclid was building a prototype 340-tonner with a Siemens AC-current electric traction system in 1999. To test the principal, Euclid had fitted one of its 280-ton Euclid R-280s with the AC-current system and it was tried out in 1998 at a coalmine in eastern Pennsylvania.

Late in 1998, Liebherr Mining Equipment introduced the prototype of the gigantic T-282 dump truck. This 380-ton capacity hauler was originally designated the KL-2680 in its post-Wiseda incarnation, and incorporated a Siemens-Liebherr AC electric traction system along with its main 2650bhp MTU/DDC 16V-4000 diesel powerplant. The T-282 was shod with 55/80R83 radials on 210cm (83in) rims,

enhancing its tall stance. This heavyweight registers 536,599kg (1,183,000lb) fully laden, measures 14.5m (47ft 8in) by 8.7m (28ft 7in), and was in operation at the Thunder Basin Coal Company's Black Thunder mine near the town of Wright, Wyoming in 1998.

DUMP TRUCKS IN REAL LIFE

In Europe of the 1960s there weren't as many medium-sized dump trucks as are used in the 1990s, and very few large ones at all. As we've seen, most of the big stuff was to be found in the USA, Australia and South Africa, working in the large open-cast coal and ore mines. In the UK in the 1960s the construction industry was served by fairly primitive Aveling Barfords with leaf-springs, while long established firms like AEC and Foden made off-road six-wheel dumpers that were really glorified trucks. Volvo brought in the first six-wheel articulated dumper, which, though it was excellent through the worst of terrain, was painfully slow on good going.

According to Stubbsy, the only truck to be taken seriously on good haul roads was the Caterpillar 769. This moved relatively large loads quickly, but was useless in bad conditions as it was two-wheel drive. In those days the hydraulic excavator wasn't used for bulk loading because the machines just weren't around, except in the very largest sizes. Dump trucks were loaded by dragline or wheeled or tracked loading shovel.

Things changed somewhat with the acceptance of the new generation of hydraulic excavators as prime-movers, and loading shovels fell out of use in this application except for loading loose materials, but they were no longer used for digging as the excavators proved themselves to be far more efficient. A typical excavator would be a Cat 235 or Cat 245 on road construction. Much larger ones were used on open-cast sites.

The Volvo articulated dumper heralded a change of direction. Moxy also produced similar machines, though the Moxy had faults that the Volvo did not.

The main problem with the original ones was the lack of speed, although when subsequently uprated, they became the norm for the industry. The basic configuration was copied by firms such as Heathfield Haulamatic and DJB, who made a dumper that was substantially larger in capacity than the Volvo, and the same vehicle was available in larger sizes. They used Caterpillar powertrains and axles in their machines, and one DJB machine directly challenged the Cat 769, and was superior to it in that it would plough through really heavy going that would stop a 769. Caterpillar were obliged to buy DJB out, and represent them as part of the Cat range.

The four-wheel drive Cat 769, Cat 773 and Cat 777 are now confined to mining work and open-cast quarries, although you can occasionally see Cat 769s involved in UK motorway construction, along with the Aveling Barford RD65. The Cat 777 is needed to accommodate the larger excavators, such as those used at Stobb's Wood open-cast mine in Northumberland, where, incidentally Stubbsy was involved with the preliminary top soil strip. He recalled that one of the exca-

Drivers may compensate for long and unsociable working hours with hedonistic excess during brief leisure periods. The driver of this Cat 769 dump truck may have simply fallen asleep while crossing the 24m- (80ft-) high Bailey bridge during construction work at Bristol West Docks.

vators was an O&K face-shovel, which arrived from the factory on board a fleet of twelve exceptionally large low-loaders. It took the engineers three weeks to assemble, with the aid of two massive telescopic cranes. Even the brand-new Cat 777s arrived in four basic sections and were assembled on site. These machines are there for life, as once they're built up, there's no way of getting them out again unless they're dismantled. A normal road wouldn't accommodate a 777.

When Stubbsy was working on the Chelmsford by-pass in the early 1980s, he recalls there were a few Russian-made dumpers there. They were imported by UMO Plant, and bought fairly cheaply by the contractor. They were 6x6 and looked like they'd been produced fifty years earlier. In order to get them to tip, the driver first had to put the box into neutral. He then had

A Cat 936E wheeled loader charges dump trucks at a quarry where material is dispensed from a rail track hopper. When one truck is full, the next reverses up to the loading point. If it takes more than five passes to fill the dump truck, the operator isn't using big enough loading shovels.

to engage the power-take-off with something that looked like a handbrake lever. Fourth gear on the main box then had to be engaged, after which another handbrake lever had to be engaged to make the connection between the p.t.o. and the pump.

The first thing that strikes you about a Cat 769 is its size. It's a four-wheel dumper with single wheels on the front axle and twin wheels on the rear. There is in fact no front axle as the wheels are mounted on a type of MacPherson strut that's basically a nitrogen ram bolted to the chassis, which allows the wheel to travel vertically. That's all there is. The wheel hub is bolted to the bottom of the ram, and it steers by means of hydraulic rams. There is an axle at the rear, however. This is suspended by two nitrogen cylinders located either side of the differential and attached to the chassis.

The chassis is a simple ladder-type construction, and the cab is on the left-hand side of the machine, right at the front. To the right of the cab is a deck with an engine cover in the middle, and the batteries in a carrier at the far side. There's a door onto the deck, and another that opens onto a catwalk that leads round the front of the machine to a ladder that gets the driver up and down. The dump body is hinged at the back and is raised by two hydraulic rams. The exhaust from the engine mates up with the body when it's down and flows round the inside of the double skin of the body, heating it up before exiting from the rear of the body. This is to stop material from sticking to the body and building up deposits inside it. There's a scow-front or shield to the dump body, which protects the cab and deck, and when fully-loaded, part of the

DUMP TRUCKS

load will be carried by the scow-front.

Driving the vehicle in reverse is easy, as it is fitted with excellent mirrors, and you can see the ground behind the wheels. Driving it forward is more tricky, as there's a lot of dead ground in front of the vehicle, to the extent that small vehicles on site are often fitted with telescopic masts carrying flashing lights, so that they can be seen by dumper drivers. Says Stubbsy:

'A frequent accident used to be caused by an engineer driving up the blind side of a 769 while the driver was waiting his turn to be loaded and probably reading his newspaper. The engineer would park in front of the dumper, unaware that the dumper driver couldn't see him, and hadn't seen him arrive. It wouldn't occur to the engineer that something as substantial as a Land Rover was invisible. The dumper driver would suddenly realise that the loader was waiting for him, whack it into gear and give it the boot, wondering a second later what the crunching noise was. As he turned to reverse to the loader he would observe people running towards something behind him, and a heap of crushed aluminium on the ground. On getting out to investigate what had happened, he would find a two-foot high Land Rover jammed under the differential of his truck, and if he was lucky, a live engineer.

DUMP THE DETRITUS

Here's a driving lesson from Stubbsy in a Cat 769 C dump truck. First thing is to stick into top gear – 8th – press the throttle and away she goes. You change gear, up and down, according to ground conditions. It also has a retarder that acts on the torque converter hydraulically, thus braking the engine, and thus slowing up the vehicle. This is used to reduce its rate of progress down ramps and long descents, and saving the brakes as well. In some conditions, if the dumper was allowed to free-fall down a long incline when loaded it might gain sufficient impetus to overspeed the engine. This normally results in a lot of bent valves. The retarder prevents you from getting into that situation, and is also a back-up in the case of brake failure.

Now it's your turn to back up to the excavator. While the previous truck was being loaded you positioned yours to give you a view of the excavator in your mirrors. As soon as the other truck moves off, you reverse quickly, keeping the truck in line with the excavator. When you hear a crash in the dump body and the truck jumps, you stop. You should know how many cycles the excavator will take to load you. It might be three, four or five, maybe six. If it is six, then they haven't got a big enough machine loading you and they are losing money! In any case the excavator will probably sound his horn when he's dumping the last bucket.

Before the last bucket lands, you'll have put your vehicle into gear and be holding it on the foot brake, so that you get away swiftly. Off you go down the haul road, keeping to the left (assuming we're in the UK). Half-way down the road you see a grader working the haul road, which is approaching you on your side of the road. There's also a dumper coming towards you on the other side of the road, his left. The grader-driver seems unaware that there is a hazardous situation developing. This is because the grader, when working the haul-road, has absolute right of way and doesn't have to deviate from his chosen path for anything or anyone. It's therefore up to you to circumnavigate him. In this case, the truck coming towards you will tuck in behind the grader, and you will cross to the other side. This way there's no collision and, more to the point, no-one has to stop.

When you get to the fill area or tip, you reverse in where directed by the man on the ground or just use your initiative if no-one's there. You tip the load by grasping the tipping lever to the left of your seat and pulling it back. Forty tons of material and the weight of the tipper body itself rise into the air in the space of a couple of seconds. You move forward, the lever having jumped forward into hold as you let it go. You now push it down one position and the body is powered down hydraulically. When it's down, you push the lever into its last position, which is float. Then you go back to the excavator, and so on until the shift ends. You will find, says Stubbsy, that it can quickly become quite tedious, as there's no creativity in it, and the monotony is difficult to overcome. An existential novel like Sartre's *Road to Freedom*, read in snatches, will give you something to think about until you get back to the digger and can read another paragraph while you wait. As for driving ability, you only need a talent for reading the ground conditions.

CRANES

A visit to a major construction area such as London's Docklands will demonstrate just how vital the crane is in the construction industry. In what used to be the old Port of London, vast new buildings of bizarre architectural design were being erected at feverish pace from the mid-1980s and, at the turn of the new millennium, were still going strong. The forest of cranes involved in these projects was breathtaking, and you wondered how they managed not to bump into one another.

The handful of conserved dockside cranes, standing like proud sentinels as a monument to the former age of shipping and bustling commerce, were dwarfed by the spindly tower cranes that soared above them from among the sleek, glass skyscrapers emerging on the former wharves and quaysides. In some of these watery enclaves, houseboats and yachts have taken the place of freighters and cargo ships. You can see cranes just like the Docklands versions – perhaps working individually – at building sites in virtually any town and city. They're all erected on site and they stay there until the building work is completed.

The principle of the crane is relatively straightforward: it is basically a powered lever. This concept was well used in antiquity, when the motive power was provided by human or animal muscle. The stones forming the ancient monuments like Stonehenge and the Pyramids of Egypt were undoubtedly raised by levers and pushed along on rollers, and probably moved into place on ramps. As an example

Cranes are a major feature of any construction site, such as the Rapier (left) seen here at the Channel Tunnel beach-head on the English Kent coast, hoisting steel piles into position for the creation of a harbour terminal.

of how they performed such tasks prior to the advent of powered cranes, a detailed illustration in the Vatican library shows how contemporary civil engineers erected an obelisk in 1585, brought to Rome from Heliopolis in AD 40 by Emperor Caligula. The architect and engineer Domenico Fontana was hired by the Pope to mastermind the removal of the 22.8m (75ft), 270-ton stone to its new location in front of St Peter's cathedral. He utilised an enormous wooden scaffold and 40 capstan winches and levers, plus a host of ropes and pulleys hauled by over 800 men and 140 horses. It took them two weeks to perform what might have been done in hours by modern cranes, given suitable planning and forethought. The manually operated windlass acting with a short jib was a significant tool in the arsenal of the architects and labourers who constructed the great European cathedrals and churches of the Middle Ages. Medieval illuminated manuscripts depict this, such as the 14th Century Solomon Temple by Jean Fouquet. Drawings of contemporary chroniclers like Matthew Paris and Vilars de Honnecourt show individual blocks of dressed stone being raised in this way during the construction of cathedrals and châteaux. In order to hoist even heavier loads, they constructed huge treadmill-like wheels on site and, in some cases, the wheel itself was hoisted aloft to serve as a giant winch.

But it's the maritime arena that takes precedence in the early evolution of the mechanical crane. The scaffold system used in the Roman obelisk erection involved the same goat-leg hoisting principle still employed aboard Dutch sailing barges – albeit on a much smaller scale – to raise and lower their masts when negotiating bridges on the inland waterways. Small hand-powered cranes were used by Chinese traders aboard their junks a lot longer ago to transfer cargoes and produce from ship to shore, and the difference was quite marked. The crane principle meant that the jib could be slewed from one side to the other, instead of simply straight up and down like the goat-legs. In this way, the commodity being transported could be hoisted from the ship's hold on to the quayside.

NAVAL BASE

It wasn't until the advent of steam power in the nineteenth century that the powered crane took off. The ready availability of wrought iron and steel enabled some extremely large cranes to be built, and among the first engineering projects to put them to use were the consolidation and expansion of naval dockyards. To carry out such tasks, large cantilever cranes were erected to enable breakwaters to be built out to sea in order to extend existing harbours. A good example of this is the Admiralty Harbour at Dover, constructed between 1898 and 1909, and extending to 1173m (3850ft). It enclosed an area of sea 244 hectares wide and 18.3m (60ft) deep. The operation was carried out by casting over 60,000 30-ton concrete blocks on site, and lifting and setting them in place with steam-powered Goliath cantilever block-setting cranes. Each of these cranes weighed 100 tons. But they were only just getting going, as manufacturers Stothert and Pitt – whose name still adorns dockside cranes – created the enormous Titan block-setting cranes which tipped the scales at 576 tons.

The UK engineering firm Stothert & Pitt, originally renowned for building early steam railway engines and mill engines, was one of the first companies to manufacture cranes. By 1850, Stothert & Pitt were making hand-operated 3-ton cranes for use in quarries, and a decade on, steam-powered cranes had taken over as their principal

The massive redevelopment that transformed London's old West India Docks into a modern city suburb with marinas in the 1990s was made possible by a forest of cranes of all descriptions. The Canary Wharf centrepiece even had cranes on the roof during the build programme.

line of business. A 6-ton steam-powered crane was unveiled at the 1867 Paris Universal Exhibition, where it won a silver medal, and other awards for Stothert & Pitt cranes followed in London in 1885 and Paris again in 1889.

The company's self-propelled Titan crane could raise up to 60 tons at a radius of 30.5m (100ft). It ran on a 5.2m- (17ft-) wide track, and its 16 wheels were mounted on four bogies. Eight of these wheels were driven from the main lifting engine by means of bevel gears and drive shafts. A typical power system for a Titan would have been a twin-cylinder steam engine, although electric-motor systems were available. It was normal for Titan cranes to be assembled on site, and sizes varied according to each crane's intended purpose. Typically, a Titan's superstructure was carried on a 12.1m (40ft) roller-track bearing, so the cantilever arm was able to rotate through a full 360 degrees. This was accomplished in about three minutes. Its legs were normally 9.1m (30ft) tall. The 61m- (200ft-) long cantilever arrangement supported at one end the operator's cab, the steam engine and its vertical boiler, plus heavy counterweights located beneath the cab to balance the weight of the object being lifted and hauled along the crane's operating arm. A trolley known as a jenny was drawn along the cantilever by control wires, and this in turn housed the multiple lifting cable that ran in four pulley wheels and from which the hook or bucket was suspended. However, because the Titan used a four-sheave block, its load could be raised and lowered only very slowly.

In the main, Goliath and Titan cranes were employed in the construction of harbour facilities, where cranes proliferated from the 1890s up to the 1960s. In the mid-1960s, the combination of short-sighted dock strikes and the onset of containerisation spelled the end of most of the UK's traditional cargo ports and changed the face of shipping the world over. Containers and the automation that came with them, of course, meant that the services of the manual dock-worker were no longer required, although crane-operators were in demand to lift the container from truck to ship's deck and vice versa. Containers were all about uniformity. There was no way of knowing what commodity lay inside one, unlike the days of the crane when dockers could see what was suspended from the hook. Containers were lifted by a specialised type of crane that stacked them on or off specially designed container ships.

One of Stothert & Pitt's early creations, built to the design of William Fairburn in 1876, is preserved in working order at Bristol docks. With a lifting capability of 35 tons – at a time when most could only handle 3 tons – it had an arched jib made of wrought iron,

The growth of containerised cargoes and the consequent need for vast bridge cranes like this Demag 20-ton lift model largely brought about the demise of the old dockside cranes (above). The chief similarity between old and new was that both moved along on tracks.

Below: Specialised Caillard container-handling cranes unload the *DSR America* alongside Bougainville Quay in Le Havre docks, Normandy. The feeder vessel being loaded in the foreground will take some of the cargo on the next leg of its journey. **Left:** underslung carriers traverse the crane jib with the containers.

and the turntable base for the 360-degree slewing mechanism was housed some 4.5m (15ft) below ground, ensuring a stable operating platform. The Fairburn design at Bristol differed from the majority of dockside cranes in that it was static, whereas most others moved on rail tracks. Dockside cranes made by the two leading companies Stothert & Pitt and Mather & Platt were powered by electricity and ran on rails along the length of the quayside. Power for the electric wheel motors was provided by long cables that snaked across the quay as the cranes travelled along. Usually, up to four cranes worked alongside the cargo holds of the vessel being loaded or

CRANES

The crane mounted on this barge – the Hugh Gordon – is used in off-shore pipe-laying at oil and gas rigs and terminals. The vessel was built in 1979 by Brown & Root of Houston, Texas, the largest engineering construction firm in the USA, using NASA-derived computer design.

unloaded, and each dockside crane's capacity was about 23 tons. In general, their format was fairly similar. Like miniaturised Eiffel Towers, they rose from their tracks on a latticework of four slender girders to a height of about 9.1m (30ft), surmounted by the turntable platform. The turntable, electric lifting and walking motors, jib and controls were all housed within the operator's cab. Above this, the primary jib was hinged off the sides of the cabin, while at the apex of this was a shorter jib that pivoted off the main one. Behind the cab's multipane glass front worked the crane operator. From his lofty erie he could look down into the depths of the ship's hold to view whatever commodity he was unloading, guided where necessary by dockers stationed in the vessel. Once freed from the bowels of the freighter, the cargo was craned down on to the quayside or into waiting trucks or railway wagons for dispatch into wharf or warehouse store. Railroad tracks were almost inevitably placed alongside the crane's tracks for convenience, frequently taking the form of standard-gauge steam goods trains. Old Victorian warehouses usually had a small crane on the top floor, known as a jigger, and each floor of the warehouse had a doorway through which goods could be brought in or passed out. Some of the seventeenth- and eighteenth-century Dutch merchants houses in Amsterdam still display this feature.

With the container revolution, the crane became redundant in the London Docks just as it did in the UK's other major ports of Liverpool and Southampton, although Bristol's Royal Edward Dock retained a number, including the 25-ton Arrol crane of 1911. This was still in working order nearly 90 years on, and the oldest electric crane commercially operable. Cargo ships continued to use the quays at Avonmouth in the traditional manner to load or unload general cargo on into the 1990s. They were served by more modern regular dockyard cranes, manufactured in the late-1950s and early-1960s, which had replaced the old-fashioned steam and electric versions. Most popular among the crane drivers was the Stothert & Pitt Blue Streak model, which was not only modern in function, but allowed the controller to sit rather than stand to do his work.

SHIPS' DERRICKS

In the pre-container era, cargo ships were fitted with their own on-board cranes known as derricks. At its most basic, the wooden boom of a gaff-rig sailing barge could serve as a crane for raising or lowering minor cargo – I even used one once to raise a damaged lee board from the murky waters of a canal – and, at its most sophisticated, a derrick could offload a ship's cargo into lighters moored alongside if no suitable quay was available.

A derrick consisted of a steel boom supported by a wire coming from the ship's mast and pivoting at its heel by the base of the mast. Some derricks were mounted independently of the mast and pivoted within short Samson-posts. Derricks could be powered electrically by the ship's engines or their own electrical or steam generators. The term derrick also refers to a small crane positioned on top of a skyscraper, usually working in consort with a tower crane. The expression 'derricking' applies to the motion of raising or lowering a crane's load by adjusting the inclination angle of the jib in relation to the tower element of the crane.

Before leaving the nautical environment, another type of crane needs mentioning. This was the gantry or bridge crane, which was used in shipbuilding, and consisted of a latticework of girders, basically shaped like a pair of giant triangles with a crosspiece along the top. The operator's cab and hoisting tackle were mounted on a track system high up on the cross-bridge, and the gantry itself moved along tracks in the shipyard in the course of constructing the vessel.

By the 1990s, the wheel had gone round half circle. There were undoubtedly plenty of cranes operating at busy, modern container ports such as Felixtowe and the Hook of Holland. Some of them were quite highly evolved cable-operated gantry cranes. But with the redundant dockyard cranes of the Port of London remaining in place only as static tourist attractions on the quaysides of the newly gentrified Isle of Dogs, a new breed of crane dominated the skyline. The vast construction site of London's Docklands presented a curious social and architectural paradox with the post-war tenement housing of the rest of the East End, and while building work was underway, the tower crane ruled supreme. Up sprang waterside apartments and office blocks, like the towering Canary Warf, and they were visible for miles around.

The Kroll K-10000 is a hammerhead tower crane with cantilever arms, which can lift 100 tons up to 122m (400ft) high. The Kroll K-10000 spans the circumference of six soccer pitches. The rear section of the crane that carries the counterweights is longer than a Jumbo Jet.

VERTIGO SUFFERERS DON'T APPLY

Among the principal manufacturers of tower cranes in the 1990s was Beck & Pollitzer. Their impossibly spindly giants have to be constructed in situ from components transported in by means of telescopic mobile cranes which assist with the elevation of parts. There are limits to their effectiveness, as the higher the jib, the less it's capable of lifting. Quite clearly they are suscepti-

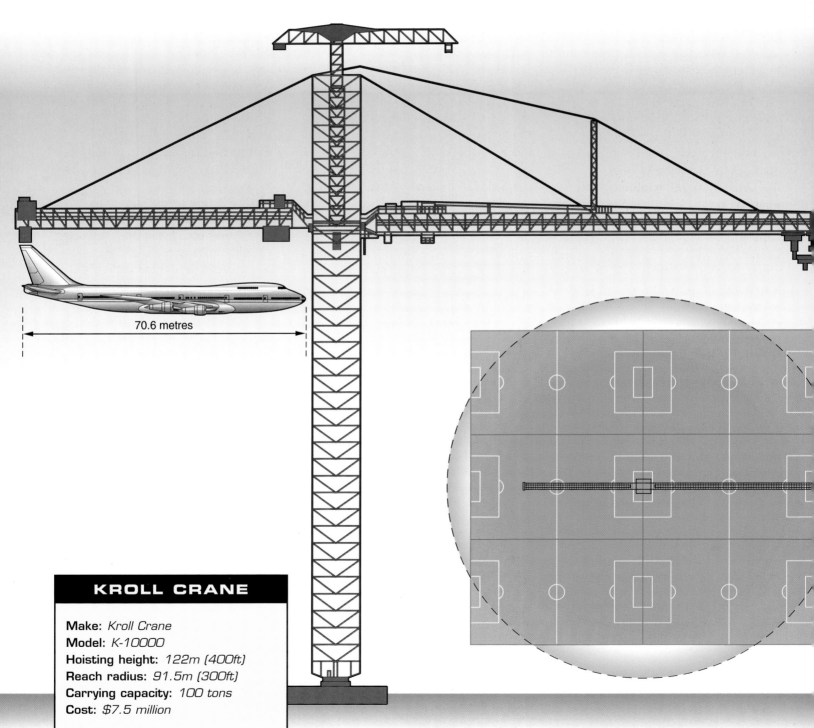

KROLL CRANE

Make: *Kroll Crane*
Model: *K-10000*
Hoisting height: *122m (400ft)*
Reach radius: *91.5m (300ft)*
Carrying capacity: *100 tons*
Cost: *$7.5 million*

ble to wind velocity, both in terms of actual crane height and the effect the wind has on the item being lifted. Consequently all tower cranes are equipped with anemometers that record prevalent wind speed so the operator can react accordingly. The height of a tower crane is balanced by proportionately lengthy outriggers that form a stable base.

You're more likely to meet tower cranes in European construction sites, and the mobile crawler variety in the USA. This has to do with environmental issues. In European cities there tends to be less available space on building sites, and the static tower crane can be anchored on site for the duration of the programme. There is therefore no need to compromise the mobility of a crawler crane. Space tends not to be so much of a factor in US cities and, in addition, the steel reinforcements frequently used in skyscrapers used to weigh more than tower cranes were capable of lifting. So the preferred option was the more robust crawler crane – which would have a similar maximum lifting capability as a tower crane – allied to a derrick sited on top of the progressing building in order to give it the extra necessary height. A mobile crane's effectiveness diminishes the higher it has to lift its load, whereas a tower crane is able to carry a heavier load at a greater radius. The major limiting factor about a tower crane is not so much its stability as the weight of the cables it has to carry. But the key to the effectiveness of a tower crane is that it can hoist aloft prefabricated components of a building, so they don't have to be assembled in situ in potentially perilous circumstances.

A typical tower crane is the hammerhead variety, which is composed of a tall lattice-work of girders forming the tower, with a crossbeam, consisting of two separate lengths, heavily braced from the top of the tower section. A system of counterweights on one side of the crossbeam imitates the passage of the winching apparatus that tracks along the other longer beam conveying the hook and its cargo.

One way of reducing the ground space and counterbalancing element needed by a tower crane is to lock the tower section into the building that's under construction, rather like an exterior lift. The jib remains clear of the skyscraper as it progresses upwards, free to hoist further sections of masonry or mortar. This was one method used at London's Canary Wharf development during the 1990s.

While some tower cranes are small 6-m (20-ft) beam models with a 1-ton lifting capacity, and are often used in inner city building sites and urban developments, the largest tower crane in existence in the late 1990s was probably the Kroll K-10000. This was a soaring hammerhead tower crane with cantilever arms, capable of hoisting 100 tons to a height of 122m (400ft) in the air, and reaching more than a 91m (300ft) radius. As if to gild the lily, it was even surmounted by an auxiliary crane that serviced the main crane and its arms and lifting tackle, and which itself had a reach of 30.5m (100ft), which was no mean feat. The Kroll K-10000 was built up on a square base with 28 feet locked into each side. It was employed on civil engineering contracts, including power stations, offshore drilling rigs, and chemical plants.

The downside of the cloud-pulling Kroll was that it was so expensive to erect and then dismantle and take away at the end of the project. In the mid-1990s, it cost 7.5 million dollars to buy, and required no fewer than 100 semi-trailers to transport the separate sections of lattice-work it was comprised of.

RAILWAY CRANES

Railway breakdown cranes played a significant part in the evolution of the mobile crane. Road-going motor vehicles had become sufficiently reliable so that by the 1990s, breakdown trucks were less familiar than they had been in previous decades. In any case,

Two very different cranes operating over ventilation shafts at the Anglo-French Channel Tunnel site. On the left is a crawler crane with lattice work jib, with a rail-track mounted bridge crane on the right.

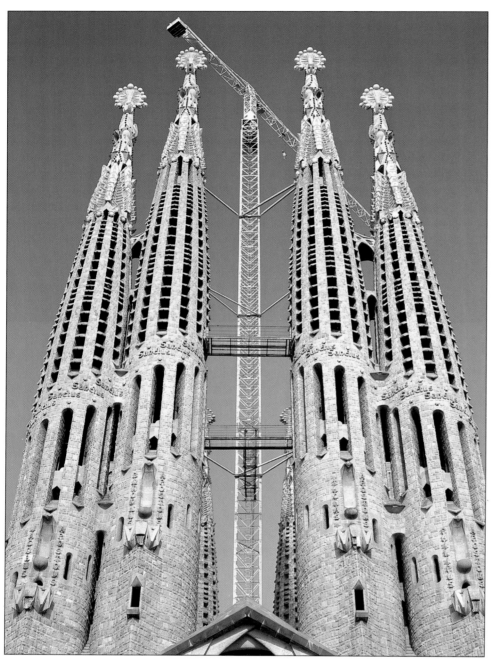

A hammerhead tower crane soars between the four fantastic openwork spires of Antoni Gaudi's Church of the Sagrada Familia in Barcelona. The faceted finials at the top are decorated with broken coloured tiles, but with a building programme extending from 1903 to 1926, this extraordinary building is as yet unfinished.

1960, steam-powered breakdown cranes were being exported to India for use on the railways. In most other countries, breakdown cranes used diesel-electric power after the disappearance of steam engines. One of the first diesel-electric applications was made by the Carlisle-based firm Cowans-Sheldon, and this was an enormous 250-ton wrecking crane built for a Canadian railway company. But in the UK, most diesel-electric breakdown cranes were built with a relatively small 25-ton capacity, and work in tandem in the event of having to sort out a serious train crash. As well as clearing the lines, more regular duties involve moving or installing track-side features such as power gantries, signal posts, bridge girders, or hoisting sections of track.

MOBILE CRANES

While dockyard and railway breakdown cranes were restricted to the tracks they served, general purpose cranes that operated in construction sites, quarries, and builders yards enjoyed greater mobility, provided the ground was completely level. Because that wasn't the case in construction sites, the crawler crane was invented in the 1930s to take care of undulating working conditions. It's no surprise to find that the first ones were developed by Caterpillar as an extension of its agricultural- and construction-site plant machinery. While they could operate quite satisfactorily within the bounds of the building site, they had to be shifted to other sites by low-loaders. The lattice-work jibs of crawler-type cranes were dismantled for transportation. However, mobile cranes went back a good deal further than that. Henry Coles and his three brothers

winches had superseded the craned tow-truck of the 1950s. But the railway crane was more heavy-duty in format. When railway trains suffered what was euphemistically known as a mishap, the breakdown crane was sent in. Originally steam-powered, the crane was mounted on a flatbed truck with standard-gauge railway bogies, and drawn by a shunting engine. Other carriages in the breakdown train included a mess coach, a few open trucks into which any debris could be placed, tool vans containing jacks and cutting gear, and timbers for propping up the wreck. The breakdown crane was capable of raising some 25 tons but was restricted by the width of its base, a flat bolster truck known as the runner, which conformed to the relatively narrow 1.4m (4ft 8in) track gauge-width. Stability was increased by extending a set of steel outriggers. The idea was that the breakdown crane would salvage the wreckage and remount the undamaged engine and rolling stock back on the tracks. Among the manufacturers of British railway cranes were Appleby, Cravens and Cowans-Sheldon, and even as late as started off working with London-based crane manufacturers Appleby Brothers in the latter part of the nineteenth century, and they teamed up to form the Henry J. Coles Company in Southwark, south-east London. By the 1890s, Coles were making fully slewing, steam-powered, railway-mounted cranes and marketing them successfully on a worldwide basis. In 1897, Coles introduced the first such crane with a hydraulically-driven jib, and by this time the firm had relocated to Derby, where its premises was called the London Crane Works. Coles's 40-ton steam crane of 1907 was the largest railway-mounted machine of its day, and its first road-going crane appeared in 1922. Invented by Arnold Hallsworth – who went on to become Coles's managing director – the 2-ton mobile crane was based on a Tilling-Stevens petrol-electric bus chassis. As with a diesel-electric set-up, the petrol engine generated the power to run the electric traction motor, as well as the crane function. By using traction-motor drive, there was no need of a clutch or gearbox.

The Coles family bowed out of the business in 1926, leaving Hallsworth at the helm to weather the great depression, with few orders taken. In 1936 the firm was on its knees, despite coming out with a 25hp diesel-powered mobile yard crane in 1936. Hallsworth tendered successfully for an Air Ministry contract to produce a mobile self-propelled crane with a 2-ton capacity and fully slewing capability. This saved the business, and Coles set about producing 82 Electric Mobile Airfield cranes. The prototype was cobbled together on a Morris Commercial chassis, with a diesel engine powering a generator that drove the traction wheels and motivated the crane mechanism. Production versions were constructed for the most part on Thornycroft truck chassis frames, some at Derby, and others at Sunderland in the Steel Cranes plant that amalgamated with Coles in 1942. The EMA cranes were built in seven different versions of up to a 6-ton capacity, and figured on a great many

Carrier-mounted cranes don't come much bigger than this. The telescopic boom expands further with a lattice jib to a height of 134m (440ft), and with stabilisers extended, it can lift a Herculean 880 tons. There are two separate engines – one for the tractor unit and one for the crane.

With all its elevating boom, jib and stabilisers telescoped away, the Liebherr LTM 1500 presents a tidy package. Its lift capacity is 500 tons, and the operational length of the boom is between 16 and 84 metres (52 and 275ft), with the lattice work jib raising the height capability to 91m (298ft).

Allied airfields during the Second World War, removing downed aircraft from runways, hoisting engines for installation, and performing sundry other duties.

Coles were onto a winner. Not only were their products used in a military capacity, but when peace resumed in 1945, they were also sought after for the business of reconstruction and reparation of war damage. In 1954, Coles offered the largest mobile crane in the world, the 41-ton Colossus, which was widely used in the construction industry. By the mid-1950s Coles was routinely offering diesel-electric mobile cranes of up to a 20-ton capacity. Then in 1959, the company was further strengthened when it took over the two smaller but innovative firms of H. Neal and F. Taylor. This gave Coles an introduction into the world of hydraulic telescopic-boom cranes that both Neal and Taylor had been exploring. The first vehicle to be produced to a hydraulic telescopic format was called the Coles-Hydra, and it proved so successful that Coles eventually marketed a wide range of the hydraulic telescopic line. Whereas a regular crane jib was fashioned from steel lattice-work, the boom of a telescopic crane was made of steel box-sections that telescoped into one another by means of hydraulic rams. This effectively quadrupled the potential length of the jib, so that an extremely tall mobile crane could be dispatched to virtually

CRANES

LIEBHERR CARRIER CRANE

Make: Liebherr
Model: LTM1500 16 x 6
Type: Hydraulic, carrier mounted
Manufactured: 1996
Engine: Daimler-Benz OM423A (crane) & OM444A (tractor)
Power output: 300bhp & 400bhp
Rated lift capacity 500 tons

any given site. It was stabilised by means of retractable hydraulic outriggers, and was also very compact, because the telescoped jib could fold up and lie within the vehicle's chassis.

Meanwhile, Coles was still producing conventional lattice-jibbed models, the largest of which was the Centurion, which appeared in 1963. Again, this was the biggest mobile crane in the world, with a lifting capacity of 100 tons. The Colossus line was relaunched in 1971, with a capability of raising 200 tons, but these cranes weren't strictly mobile in that they had to be erected on site from a kit by means of a mobile telescopic crane to raise the componentry, which took perhaps three to four days. A variation on the theme was the telescopic boom that had a lattice-jib extension, which could attain more height, while the fly-jib and luffing-fly jib pushed the envelope even higher, doubling and even trebling the operational height of a crane.

The lighter-duty mobile cranes proved more versatile for certain applications and, by the 1990s, were produced by a wide variety of manufacturers, lending themselves in the process to the crane-hire business. One leading operator in this field was Sparrow Crane Hire, which started life renting out ex-WD army surplus cranes and, by the 1990s, was the purveyor of 500-ton capacity machines on a contract hire basis. Manufacturers active in the production of lattice-boom and telescopic cranes by this time included Kato, Link-Belt, Lorain, Marchetti, Manitowoc, Mannesmann Demag, P&H, and Tadano Faun.

Another manufacturer with a broad product base was the Pennsylvania-based Grove Cranes, which started out in 1947 as an agricultural wagon maker, and came out with its first crane in 1949. It was constructed in the first place simply to render the company self-sufficient in the materials handling department. But demand from customers made it an instant success. A decade on, Grove introduced its first four-wheel drive, four-wheel steering mobile crane. By 1965, production had totalled 1000 units of various different types of crane, and production doubled in two years. This made it an attractive takeover prospect, and Grove Manufacturing was acquired by Kiddle Incorporated of New Jersey in 1967. Output continued to rise, so that by 1979, Grove had built over 20,000 cranes. That year its horizons expanded with the acquisition of the Californian Manlift Company, which made a range of complex boom and scissor aerial work platforms such as those used by electricity-pylon maintenance and window-cleaning contractors.

In 1982, Grove's products were commended by the US Government, which conferred the US Defense Department's Contractor Assessment Program Award on Grove Manufacturing. The following year it came out with a hydraulically-powered lattice-boom range. This was promptly seized upon by the OshKosh Truck Corporation which needed a specially designed materials handling crane for mounting on the rear of its spectacular Heavy Expanded Military Tactical Truck, better known as the HEMTT. Oshkosh ordered no less than 6500 cranes from Grove. The company's next move was to acquire Coles Cranes in 1984.

The US Military awarded Grove Cranes a contract worth 50 million dollars in 1986 to supply 269 rough-terrain container cranes which were designated the RT875CC. The same year, Grove's first 336-ton capacity TN3000 mobile crane was delivered to a Swiss contractor, which went down in the annals as the largest telescopic-boom mobile crane ever made in the USA. As for Coles Cranes, its custody reverted to the UK once again, as ownership of the parent Kiddle Incor-porated moved across the Atlantic to Hanson Trust plc.

In 1988, Grove brought out its AMZ66 articulating boom Manlift aerial work platform, and plans were underway for a new plant at Salem, Virginia. The company name changed to Grove Worldwide in 1990, reflecting its international status and wide range of vehicles. These included the all-terrain hydraulic units which could cruise the main roads at 80.4km/h (50 miles/h) while offering good off-road axle articulation. Grove's rough terrain cranes were built on box-section chassis frames and incorporated four-wheel steer characteristics. These machines could operate at heights of up to 82.3m (270ft). Typical of these was the truck-mounted (TM) Grove TM9120 hydraulic crane, which was a four-axle chassis with hydraulic retractable outriggers and a hoist operated by a variable displacement piston motor incorporating pressure override for infinitely variable speed. It worked in conjunction with planetary reduction gears with an automatic spring-applied disc brake. The boom was raised by means of a double-acting hydraulic cylinder with an internal holding valve that provided elevation of up to 80 degrees. The TM9120 had a lift capacity of 132 tons. The tractor itself was an 8 x 4 configuration, powered by a 320bhp Cummins N14-460 turbodiesel engine, allied to a 10-speed transmission with three reverse gears. An operator's cab was located alongside the boom, and the driver's single-seat cabin was to the front left of the engine compartment. An even bigger Grove product was the TM1500, which was a six-axle model powered by a 335bhp Cummins NTC-450C engine and carrying an hydraulic telescopic crane that could raise 165 tons. By 1999, Grove offered 11 different truck-mounted hydraulic cranes, not counting its more specialised off-roading models, as well as six smaller truck-mounted materials-handling or loader cranes, and a pair of lattice-boom hydraulic cranes. Other loader crane specialists included Effer and Heila, who produced exceptional 150-ton capacity items. Loader cranes were more common in the category of up to 40 tons, with jib lengths of 9.1–12.1m (30–40ft).

Seven cable cranes each 90m (295ft) high were erected at the Itaipu Dam site by O&K subsidiary PWH Anlagen and Pohlig-Heckel do Brazil to facilitate construction of this massive Brazilian-Paraguayan joint venture to harness the River Paraná.

CRANES

The cable cranes transported 13 cubic metres (459 cubic feet) of concrete for construction of the Itaipu hydroelectric project. With a span of 1360m (4462ft), each crane could carry 20 tons, travelled on a monorail system, and could be controlled from either the Brazilian or Paraguayan side of the dam.

Meanwhile, the Japanese firm Kato, which had been founded in the late eighteenth century, offered its mid-range NK-350E-V hydraulic crane. This was based on an 8 x 4 Mitsubishi truck chassis nearly 13.1m (43ft) long, and powered by a 221bhp Mitsubishi KJ508BLE engine. Boom length was up to 33.8m (111ft) and lift capacity was 38 tons. Kato's biggest model could lift 176 tons.

The two German makes of Liebherr and Mannesmann Demag produced very large hydraulic cranes that were only mobile when transported by complex, low-bed trailer rigs. The eight-axle Liebherr LTM1800 was one of the largest mobile cranes ever made, elevating to just under 134m (440ft) and offering a staggering lifting capacity of 881 tons. To achieve this, it combined a telescopic boom with a lattice jib, and for transportation, the boom section was removed and carried separately on a semi-trailer rig that featured a steered trailing axle. The vehicle thus managed to scrape under the legislated weight limit. The Liebherr's crane engine was a 300bhp Daimler-Benz OM423A unit, and the tractor was powered by a 400bhp Daimler-Benz OM444A engine.

Mannesmann Demag's offering in the big-time mobile crane market was the neat and compact nine-axle AC1600. Minus its counterweight, it crept in below the 13.23-tons-per-axle weight restriction. It was blessed with a lifting capacity of 881 tons, and a maximum boom length of 50m (164ft). Power for the crane came from a 412bhp six-cylinder Daimler-Benz OM443LA unit, and the tractor featured a 10-cylinder Daimler Benz OM447A developing 370bhp. For ease of operation, the AC1600's outrigger controls were placed on either side of the vehicle, and the operator's cab could be tilted as necessary to give a better view of the job in hand.

ALL-TERRAIN CRANES

This category of crane is to be found principally on off-road construction sites. In the 1990s, the major players included Grove, with 16 different models, Kato, Link Belt, Mannesmann Demag, Marchetti, the French PPM concern, Tadano-Faun, and Terex Cranes. Top of the pile was the six-axle Grove GMK 6200, which was the world's largest all-terrain hydraulic crane in the late 1990s. The 12 x 8 crane carrier was powered by a 213bhp Daimler-Benz OM444A turbodiesel engine, and the 220-ton crane also used a Daimler-Benz engine, in this case the OM447A. Its boom extended to 52.7m (173ft), and the hoist used an axial piston, variable displacement motor with planetary gears, and braking system.

The French manufacturer PPM offered a line-up of eight cranes under its Quadral banner, ranging from the ATT240, rated at 22 tons, to the 121-ton capacity ATT1190. Occupying the middle ground was the two-axle 30-ton PPM ATT290, which incorporated a swing tail that remained within the base configured by the outriggers, plus all-wheel steer facility. This four-wheel drive crane was fitted with the 118bhp Daimler-Benz OM366 power unit, and the boom length ran to 27m (89ft). While the ATT290 was a middleweight crane, the heavy-duty end of the PMM range was occupied by the ATT1190. This was a 10 x 6 five-axle crane with

All-terrain cranes can operate in fairly arduous circumstances, and range in size from the Grove GMK 2035 with a 30-m (98-ft) maximum lift height, to the 40m+ (131ft+) of the Grove GM6200, Tadano Faun ATF120.5 and Marchetti MG.80.

all-wheel steering, fitted with a 118bhp straight-six Daimler-Benz OM366A crane engine and a 320bhp V8 Daimler-Benz OM442 LA turbodiesel powering the tractor unit. The telescopic boom could be elevated hydraulically to its full 45m (147ft) in two minutes, while rated lift capacity was 121 tons.

Another crane along the same lines to the ATT1190 was Tadano Faun's AFT120-5. This was the largest vehicle in the company's six-model range, and four of its five axles provided drive over rough ground, with the front three as steering axles. When being driven on-road, transmission was three axles only.

A further advance into the wilderness invited the use of the category known as rough-terrain cranes. These vehicles were not dissimilar to all-terrain cranes, but tended to be more rugged in aspect, with appropriate suspension upgrades and greater axle articulation for off-roading. Specialists in this field included the firms Bendini, Locatelli Atlas, and Pawling & Harnischfeger (P&H).

Tadano Faun's RTF 40-3 was a good example of a rough-terrain mobile hydraulic crane. A three-axle vehicle with all-wheel drive, it could cope with some of the most difficult going in the wild, with excellent approach and departure angles of 16 and 15 degrees respectively. The telescopic boom measured 30m (98ft) fully extended, and could lift up to 44 tons. Power came from a Daimler-Benz OM401 LA engine developing 213bhp. Like PPM, Tadano Faun's standard issue colour scheme was predominantly white with red detailing.

In construction-site yellow, Grove's RT9100 rough-terrain crane was endowed with the greatest lift rating of its 14-model range, at just under 100 tons, while its boom telescoped to 34.4m (113ft). It was a more compact affair than the Tadano Faun RTF 40-3 model, being a two-axle four-wheel

GROVE ALL-TERRAIN

Make: *Grove Worldwide*
Model: *GMK 2035 All-Terrain*
Type: *Hydraulic, carrier mounted, 4x4, four-wheel steer*
Manufactured: *1998*
Engine: *Mercedes-Benz OM 906*
Power output: *205bhp*
Rated lift capacity: *35 tons*

drive vehicle. Unlike the forward-control layout of the Tadano Faun, the Grove RT9100 had a single seat cab mounted between the axles and beside the crane boom. To get it across the rough stuff, power came from a 186bhp Cummins 6CTA 8.3, with six forward and six reverse speeds. The hydraulic stabiliser jacks raised the entire vehicle off the ground for operation, and could be adjusted as appropriate for uneven territory. Kato's rough terrain offerings were conceptually similar to the Grove models, and didn't provide such good angles of approach and departure on account of the projecting stabiliser and counterweight housings. The Kato KR-500, for example, embodied angles of 22 degrees at the front and 18 degrees at the rear. This was the largest of Kato's three such models, and had a rated lift capacity of 55 tons. Its boom

rose to 33.5m (110ft), with a fly-jib that could be set in one of three positions, 5, 17, or 30 degrees, and an additional luffing-jib that increased its height by a further 19m (62ft). The 4 x 4 KR-500 was equipped with the 199bhp Mitsubishi 6D22T engine which, where conditions permitted, could muster 51.5km/h (32 miles/h).

CRAWLER CRANES

Crawler cranes are more readily associated with draglines than the self-propelled telescopic boom variety. They consist basically of the caterpillar-tracked undercarriage, with a cabin superstructure housing the crane engine, counterweight, and controls. This is connected to the carrier chassis

**Left: On the River Thames alongside the old Fresh Wharf, Stothert & Pitt gantry cranes discharge freighters. The ships' derricks contribute to the unloading procedure.
Above: The modern way involves Krupp KL40 standard deck cranes aboard the cargo vessel *Nedlloyd San José*. These electro-hydraulic cranes have a capacity of 25-60 tonnes and 20-36m (65-118ft) outreach. Some ships have two cranes mounted on a common slewing platform, or double-joint and gantry cranes for container work.**

A Liebherr LR 1250 hydraulic crawler crane in operation at a construction site in Amsterdam. The lattice work jib is balanced under tension by a pair of derricks, enabling greater lift height as the cable lowers a cement hopper into place for filling up concrete shuttering. Inset: The same wide-track Liebherr model at work in Green Bay, Wisconsin. The jibs of crawler cranes are dismantled for transportation.

by a turntable or slewing ring. They have lattice-work jibs and are mainly cable-operated, hence the dragline connection. Crawler cranes are the strongest mobile machines, and although they sacrifice mobility for lifting ability, they are capable of operating in harsher environments than even rough-terrain cranes, thanks to their broad crawler tracks. They often work on construction sites in concert with tower cranes and, like their loftier cousins, have to be dismantled for transportation to and from site.

Some of the key makes in the field in the 1990s were Grove, Koehring, Liebherr, Link Belt, Manitowoc, Mannesmann Demag, P&H, PPM, RB and Sumitomo. Mannesmann Demag and the Dutch crane makers Van Seumeren both introduced an extension system that placed a lattice-work luffing beam atop the main jib, braced by a crane mast, and this was developed further. Some of the modular jib config-

The erection of the Millennium Eye ferris wheel on the south bank of London's River Thames was accomplished by a battery of cranes including this gigantic barge-mounted unit, Taklift 3.

urations that featured on the bigger modern machines were almost inexplicably complex and certainly bewildering to the uninitiated observer. There were up to 10 varying combinations of boom, depending on the application. The most highly evolved layout consisted of main jib and a second auxiliary lattice-work mast that furnished tension for the main jib via the counterbalance weights, plus a pair of lattice-work arms that provided pulley-wheels for the main jib. On some cranes, the counterweights were mounted on a separate carrier. While counterweights were an essential part of a crane's componentry, clearly they didn't want to be so heavy that they caused the crane to topple backwards. Maximum stability was gained by spreading the machine's weight as widely and as evenly as possible. One way of more than doubling the effectiveness of a mobile crane such as the Manitowoc 888 Series-2 was to incorporate a ringer attachment, which had the effect of creating a more stable base, at the expense of mobility but with a gain in altitude ability. The rated lift capacity of this crane rose from 230 tons up to 600 tons. Despite these measures, it shouldn't be taken for granted that cranes are especially safe implements, as contractors regard them as being potentially the most dangerous machine on a construction site. A tower crane operating at full capacity would cause untold harm if it should collapse in an urban environment. Cranes have to conform to a strict set of internationally recognised regulations, based on calculations governing the physics of elevating loads at particular heights and angles, low centres of gravity, and not forgetting margins of stability.

The Liebherr concern produced a range of crawler cranes with lifting capacities from 38 tons to 132 tons, and were typically to be found in construction sites, bulk-handling situations and demolition scenarios. They could also be used as draglines and clamshell excavators, fitted with cable-operated buckets and grab jaws. The Liebherr LR11200 crawler crane featured a 97.5m (320ft) lattice-jib and a rated lift capacity of 440 tons, with power coming from a 750bhp 12-cylinder Cummins KTA 38C 1050 diesel engine. It worked with a counterbalance weighing 394 tons, and the machine's operational weight was 2362 tons. Transmission was variable through two hydraulic planetary geardrives per crawler carrier. It used six winches to haul over 6.4km (4 miles) worth of cable, and its crawler pads were almost 2.4m (8ft) wide.

SPIDERS' WEBS

The German construction equipment firm of Mannesmann Demag Baumaschinen made crawler cranes with some of the most intricate jib arrangements, with cable systems as complicated as spiders' webs. They were high-lift capacity vehicles with long outreach capability, with a number of modular boom configurations going by the name of Ringlift, Light Superlift, Heavy Superlift, and Standard Mainboom. They were also interchangeable, with the format depending on the task to be addressed. The crawlers, too, were mounted on sideframes and could be dismantled for transportation. The Mannesmann Demag CC2600 was powered by a 350bhp Daimler-Benz OM442 LA diesel engine, and had a rated lift capacity of 496 tons and a maximum boom length of 90m (295ft). The track gauge was 6.7m (22ft), and the machine was rigged with twin pairs of counterweights for stability. Top of the range was the Superlifter Demag CC12600, which was also the largest crawler crane in the world in the late 1990s. Four boom combinations were available for this model, which had a lift capacity rated at 1047 tons and a maximum jib length of 233.7m (767ft). It utilised no less than 6.4km (4 miles) of cable in full operational trim. A Cummins KTA 38 C-1050 engine developing 1215bhp powered it, and the expression 'crawler' was apt in more ways than one, as it accomplished 0.5km/h (0.3 miles/h). Weighing in at 66 tons, it was perhaps easy to see why. However, when dismantled for transportation, its jibs packed away into a space just 3.9m (13ft) long by 3.04m (10ft) high.

One thing's certain. As long as we want to place objects of one sort or another on a higher level, be it cargo or building materials, there will always be a place for cranes of some description. By the turn of the millennium, there were some clever devices about, and as materials and technologies increase in sophistication, they will get cleverer still.

ROAD ROLLERS, COMPACTORS & TARMAC LAYERS

The first serious road builders were the Romans, who were able to build their routes as straight as they liked without the impediment of legislative restrictions because they were top nation at the time. This meant that they could go straight through existing settlements and plantations if need be, and only deviated slightly when faced with insurmountable escarpments or swamps.

For a taste of Roman determination, just see the way that Hadrian's Wall in northern England climbs inexorably over the rocky escarpments of Northumbria. Roman curves didn't need to be sharp, because their wagons didn't have steering axles – amazingly, this facility was not discovered until the 1500s.

Once the preparation work has been completed, the pavers and finishers come on site to lay the final surface in this case a six-lane highway. Also present is a fleet of Oshkosh S-series concrete carriers on hand to supply the giant Gomaco PS-2600, GP-2600 and T/C-600 paving equipment.

Roman roads were commonly about 4.6m (15ft) wide, with paved vehicular roads measuring on average 1.32m (4ft 8in), which is virtually the same width as a standard gauge railway line.

It's worth comparing the basic methodology of Roman road construction with that of the late twentieth century because there were marked similarities. As is the case in the modern world, the Roman roads were meticulously planned and, once approved, marked out for the soldiers from the relevant local garrisons that built them. As is the case today when a new motorway is to be built, the surveyed ground of 2000 years ago was cleared of vegetation, followed by the appropriate cut and fill operations. The stony foundations, called the pavimentum, were laid, with a layer known as the statumentum coming next. This was composed of stones bonded with mortar, and was followed by the ruderato, which was the same but with smaller bonded stones. The nucleus layer consisted of crushed stone, lime and sand – effectively a kind of concrete. The top layer was usually made of hard-wearing cobble stones. In cross-section, the roads had a pronounced camber for drainage, and kerbs and drainage ditches where necessary. The collapse of the Roman Empire in the fifth century AD prefaced the decline of the English road system, as, like Hadrian's Wall that divided England from Scotland, the locals pilfered the stone to build their own dwellings.

For a thousand years the road systems of Europe languished, although clearly people travelled far and wide. It was just that no-one made any attempt to maintain the highways. First of the roadbuilders who put matters to rights in Britain were Thomas Telford (1757-1834) and John Loudon McAdam (1756-1836), both of whom were engineers inspired by the appearance of the Royal Mail coach service that was instigated in 1784. Creator of the Menai Bridge connecting Anglesey with mainland Wales, Telford was the architect of nearly 1693km (1000 miles) of new or resurfaced roads. Acknowledging Roman precedents, he stipulated thorough foundations, and his coach roads were hailed as an engineering triumph at the time.

The same philosophy was followed by McAdam, who insisted on proper cambers and a weatherproof surface. This became his trademark, and his name endures in the tarmacadam (or tarmac) used today. It was originally achieved by compacting a layer of small stones a foot deep on the sub-strata. The regular

Diagram showing the workings of an early Aveling & Porter diesel roller, rated at 3 tons. The firm introduced its first diesel-powered model in 1927, and had pioneered the development of steamrollers back in the 1860s.

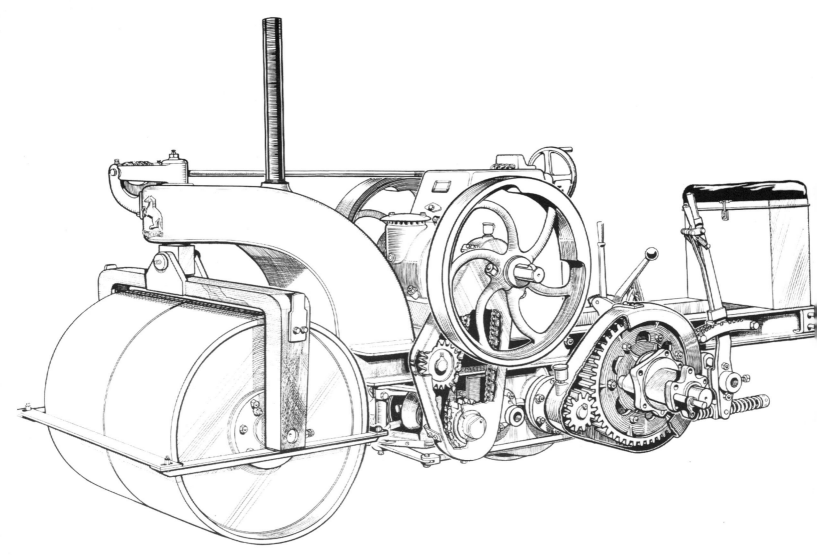

pressure of wagon wheels consolidated this process, and for a century this McAdam-inspired surface worked well enough for coaches and horses. By then, the British railway system had developed to the extent that it had superseded the roads as the main artery for conveying goods, people and produce over long distances. One reason for this was that although a certain amount of the ancient road system was resurfaced by the likes of Telford and McAdam, the network itself remained sterile, with no new road schemes being created to add to the infrastructure. Although railways very often mimicked the road system, they also struck out on fresh routes. Not until the development of the burgeoning motorway network and the construction of by-passes did major new road systems come into being. The new housing estates, overspill towns and ribbon developments that began to appear in the 1920s and 1930s clearly spawned their own road systems, albeit on a smaller scale. Greater traffic density also made a valid case for road widening programmes.

Reverting to Victorian England, inner city traffic made short work of some of the newly instituted road surfaces. Many streets were paved with cobblestones or macadam, but those made of tar-covered wooden blocks were quickly churned into a quagmire by iron-shod cartwheels. The method of construction was akin to building a wall. Courses of stone sets or wooden blocks were laid along the designated road bed, pummelled into a level surface with wooden rams, then tar was poured into the cracks as a bonding, or, in the case of the wooden blocks, all over them. Those roads surfaced by the macadam method were levelled by rollers that could be either hand-drawn or pulled by horses.

ADVENT OF THE STEAMROLLER

Then in 1862 the first mobile road-building vehicle arrived on the scene. It was the steamroller. One of the first versions was built by Aveling & Porter, but the prototype weighed in at 30 tons, and this proved to be much too heavy because of the 2237 litres (500 gallons) of water it carried as ballast. It was stored in a large tank over the single-wheel rear axle, and the vehicle was steered by means of a ship's wheel. A revised machine was introduced in

AVELING-BARFORD

Make: *Aveling-Barford*
Model: *HDC 12*
Type: *diesel roller*
Manufactured: *1999*
Engine: *4-cylinder Perkins 4236cc*
Power output: *73bhp*
Operating weight: *10,352kg (22,821lb) with ballast*

Cutaway of Aveling-Barford's HDC 12 diesel roller demonstrates how little the overall layout has changed from the original of 1927. The four-cylinder Perkins engine is located centrally for optimum weight distribution with transmission and PTO at the rear, and the steering arm is virtually direct.

1873, which weighed a more realistic 10 tons, and had tow wheels at the back and a single wheel at the front. A cast bracket bolted to the boiler casing supported the front roller, and the roller was split in two to produce the effect of a differential when it was turning. It thus set the style for steamrollers for the next 50 years or so. Aveling & Porter produced around 8800 of these units, known as the Invicta and sporting a rampant horse logo, and possibly no item of heavy plant endured for so long. The steamroller was basic and reliable and was to be found on the construction of the UK motorway network in the 1960s.

The British steamroller was the inspiration of Thomas Aveling, who was a Kent-based farmer and engineer. He developed the concept from a portable threshing steam engine, realising that this horse-drawn vehicle might actually be self-propelled. The manufacturers of the portable steam engine, Clayton & Shuttleworth, agreed to build a prototype, and this was undergoing tests in 1862. The same year, Aveling teamed up with Richard Porter and set up shop at Rochester, Kent, and the 30-tonner appeared three years later.

Following the example of railway engine manufacturers, Aveling & Porter added a second cylinder to produce a more efficient twin-cylinder compound engine. This made use of the steam a second time – firstly in a high-pressure cylinder, exhausting into a second low-pressure cylinder, which exhausted into a blast pipe at the bottom of the funnel, which made the fire draw.

The Invicta was so successful that other traction engine manufacturers came out with similar designs. Among these were Fowler, Burrell & Marshall, and Clayton & Shuttleworth, and their presence altered the way roads were resurfaced. The top crust of the macadam could be compacted more efficiently than ever before, and the steamroller played a part in the preparation of the work by scarifying or breaking up the surface of the worn-out road. The ensuing rubble was removed manually, and a layer of small stones laid down in the bed, which was crushed by the steamroller. A new crust was created by further compacting a layer of mud, which when dry was sufficiently robust to take the weight and pace of horses and carts.

SUCCESSFUL PITCH

However, the new surface wasn't up to the rigours of motorised transport, and in the early 1900s bitumen was introduced as a new binding medium. Mixed with lime, sand and aggregate, this ancient type of pitch was immediately effective as a top coating for roads. The technique, or rather the recipe, was further refined by the discovery in Trinidad in the West Indies of lakes of asphalt, which is a kind of bitumen. When heated, the mix of asphalt, lime and sand could be poured into the road base where it would solidify enough to be levelled by a steamroller. An advance on asphalt was the use of liquid coal tar, obtained by heating up coal to 1000 degrees centigrade in an oxygen-free environment. Tar is actually one of the by-products of this process, and it flows out as a black liquid. Having solidified, it was reheated in the road-building process, and was poured over the waiting macadam compound. When stone chippings were spread on the still-soft surface, and then subjected to the attentions of the steamroller, a resilient, waterproof surface was created, and this became known as tarmacadam, or tarmac for short. Originally broken up by hand, the stone chippings were produced from the 1920s by steam-driven crushers. The tarmac was susceptible to strong sunlight, which could melt the road surface, but it was sufficiently acceptable to be used for surfacing rural roads in the 1990s.

When roads needed repairing in the 1920s and 1930s, the coal-fired tar boiler and diesel-engined mixer boiler that blended the asphalt or tar with lime and sand became a regular spectacle wherever the works were taking place. The molten tar was laid by hand from oversize watering cans and spread by workmen with brooms. Stone chippings were spread over the top with hand shovels, and the surface was compacted by the steamroller. The workman's job of demolishing the old road surface in readiness for rebuilding was made easier by the innovation of the pneumatic drill, frequently made by the Broomwade company, but invariably deafening. With the new roads and

Like elephants with canopies, heavyweight Aveling-Barford rollers dominate a street scene in Delhi. The Third World condition that typifies much of India's road network means their contribution to its improvement is likely to be transitory, or at best, an ongoing presence.

ROAD ROLLERS, COMPACTORS & TARMAC LAYERS

The driver of this Dynapac self-propelled compacting roller is protected from cascading debris by a roof that also offers a modicum of roll-over protection. Also known as a sheep's-foot roller, the principal of this system is to compact the earth into itself by incessant pounding from the blunt nodules on the forward roller.

road-widening schemes instituted in the 1920s and 1930s, a different type of vehicle started to appear. This was the steam-powered tar tanker, usually a Fowler or Sentinel, the chassis of which was fitted with a tanker body. Steam was passed through coils within the tank to keep the tar hot and fluid at around 180 degrees centigrade. The molten tar was dispensed onto the prepared road bed via spray heads on a bar mounted at the rear of the vehicle. The depth of the tarmac was dependent on the speed of the tar tanker: the faster it went, the thinner the layer of tar would be. The stone chippings for the top dressing needed to be in plentiful supply, and the steamroller finished the job.

LONG DAY

The drivers of these vehicles were skilled operators, who knew exactly how quickly to go to produce the desired quantity of tarmac. The steamroller drivers had a longer day than their manual workmates, as they needed to get the vehicle's steam up prior to the start of the working day, which was generally 7.30am. They frequently had to live in primitive, austerely equipped caravans made of corrugated iron that were towed behind the steamroller. Like the main vehicle, these 'living vans' ran on steel wheels.

Another dressing medium arrived on the scene in a big way in the 1930s, which consisted of the crushed slag that was a by-product of iron and steel works, and which was treated with tar. This unlikely mixture was delivered to road-building projects and dumped in steaming heaps for men armed with multi-pronged pitchforks to spread where required. Once again, the steamroller's role was to compress and level the surface. The tarred slag was branded as Tarmac, having been 'invented' by Nottinghamshire County surveyor E. Purnell Hooley back in 1901. He made the observation that slag applied to soak up a tar spillage outside a foundry had set to a solid, flat surface, and registered the formula under the Tarmac brand name. However, the Wolverhampton steelworks magnate and MP Sir Alfred Hickman bought the Tarmac patent from Hooley and launched the product on the market in 1905. It was much in demand, especially during the First Word War for creating usable roads to the front. During the ensuing years, Tarmac Ltd colonised iron and steel works' slag heaps, establishing depots from which to dispense the product.

By this time most local county councils possessed steamrollers, possibly as many as 100 units in some cases. Professional contractors ranged from one-man-band operations to multi-vehicle companies which, by 1930, were starting to offer diesel-powered rollers. As with the trucking industry, which only began to fit diesel engines at this time, road rollers had used paraffin- and petrol-burning internal combustion engines, as well as steam, from the turn of the century. Around 1902, Barford & Perkins came out with a petrol-fuelled three-ton motor roller for use on sports fields. The same firm introduced the first diesel roller in 1927, and although it was more efficient because the driver didn't have to go through the onerous task of firing it up some hours before it was needed, it wasn't necessarily as enduring as a sturdy steamroller. While a diesel roller would require regular routine maintenance of its injectors and fuel pumps, the steam-driven machine went on rolling for perhaps ten years without needing attention. However, improved diesel technology coupled with increasingly costly coal and the necessity of a water source close to the site meant that steamrollers were eventually overtaken by diesels.

During the 20-year inter-war period, many new roads were constructed in Europe – such as the German autobahn network – and in the USA, although it wasn't until the 1950s that an intense programme of new road-building took off in the States, which included both inter-state highways and private toll-roads. In Britain there were one or two new 'garden cities', built on green-field sites, such as Welwyn in Hertfordshire. Otherwise, these new roads consisted largely of by-passes around cities to relieve urban congestion, and a prime example was London's North and South Circular roads. As a small child I lived on such a road in Leeds in the 1950s. Now considered an inner ring road, it was on the periphery of the city then, and there was minimal traffic on its dual carriageways. At night it was lit up by the new sodium vapour lights that cast an ethereal pale orange glow.

This road was eventually resurfaced with concrete. Back in the 1930s, concrete was novel in Britain, but was used extensively in the USA and in Germany for its national autobahn network. As is widely known, the recipe for concrete is sand, shingle of varying grades, and cement and sometimes lime. It sets rock-hard through a chemical reaction, and is waterproof and durable, as well as being relatively inexpensive to produce and to lay. The road-laying technique is also simple. The individual materials are blended in a cement mixer to a particular ratio, and the wet concrete is poured into the prepared roadbed that's delineated by wooden shuttering. There are ways of tamping it down to achieve a level finish. In the old days, a vibrating beam driven by a petrol engine tamped the surface, leaving a ripple effect that aided tyre adhesion in wet weather.

The upkeep of country roads was the responsibility of local county councils, and this still applies in the UK. You can find marked differences of attitude towards the desirable quality and priority of road surface maintenance from one county to the next. Living in rural Perthshire, close to the border with Scotland's Central region, it was amusing in wintertime to see the Perthshire snowplough turn around at the county line, making it necessary to trudge through the snow in the Central region on foot, because the road to the south remained impassable by car. On the other hand, the road surface was less well maintained on the northern side of the divide.

LEGACY OF WAR

It wasn't just bombs and shells that changed the environment during the last world war. Ironically, road construction and the plant necessary to create it advanced dramatically during the Second World War, and in the UK this was mostly because of the presence of the US military from 1942 onwards. The British Government sought supplies of advanced US heavy plant for the construction of hundreds of airfields and access roads, and much of this was supplied under the Lend-Lease scheme. While fighter aircraft could make do with grass airstrips, heavy bombers like the Avro Lancaster and Handley-Page Halifax, or the Boeing B17 Flying Fortress and B24 Liberator, had to have runways of at least 1828.8m (2000 yards) long. Because of variable wind directions, it was necessary to build three intersecting runways, plus a perimeter track for taxiing and adjacent areas as hard standing and dispersal points. These were made of concrete or tarmac.

In the UK, new aerodromes built during the Second World War numbered 700, and of these, the majority were constructed between 1941 and 1943. The statistics for what was the most extensive building programme since the

GOMACO CORPORATION

Make: *Gomaco*
Model: *GP4000*
Type: *concrete paver*
Manufactured: *1997*
Engine: *Cummins diesel*
Power output: *450bhp*
Operating weight: *44 tons*

Victorian railways are remarkable. The arrival in the UK of the Eighth USAF in 1942 saw a new airfield start to emerge on a green-field site at the rate of one every three days. They were built in the main by British contractors using American construction equipment, with 10 or so USAF bomber airfields laid out by US Army engineers. A single American bomber air base accounted for sufficient material to construct 3218.6km (2000 miles) worth of motorway, and that meant some 133,805 cubic metres (175,000 cubic yards) of concrete, 26,755 square metres (32,000 square yards) of tarmac, plus 12.9km (eight miles) worth of access road. Each airfield cost around one million pounds in contemporary money. In a typical situation, 12.9km (eight miles) of hedges and a thousand trees would have been rooted out. Multiply that by 700 to gauge the effect on the countryside, though the majority fell into disuse when peacetime returned.

Crucially though, the techniques and equipment employed by the US engineers changed the way roads were constructed in Europe in the 1950s. Key vehicles in the American road-building arsenal were the Caterpillar D7 bulldozer, a powerful diesel-engined machine running on crawler tracks that used a system of pulleys to raise a blade which was lowered by gravity, though superseded by hydraulic rams. Motor scrapers and graders were also imported to aid the path of the runways. After the Normandy landings in 1944, American bulldozers and graders were vital in establishing air bases for Allied aircraft. A number of fields were bulldozed into one area if required, and a device known as a wobbly-wheel roller created a flat surface. This curious apparatus consisted of a large tank that was filled with water for ballast, and a chassis from which were suspended up to 13 pivoting wheels, shod with balloon tyres and mounted on overlapping axles. The effect was to make one very large roller, and when towed up and down the field, it created a serviceable runway. Wobbly-wheel rollers were also used in the Western Desert during the North African campaign. They were originally made in the USA in considerable numbers, and were produced in the UK by Pullen Engineering after the war.

ROCKY ROLLER
The Aveling Barford Company won a contract to build a diesel roller to a specification laid down by the Air Ministry. Essentially a utility item of lightweight construction – in order to conserve steel – the Aveling Barford diesel roller featured twin tandem rollers that were filled with concrete to bring the weight back up to a realistic level. The company ensured that all the castings for the chassis, engine and gearbox were embossed with the legend 'War Design' lest anyone mistake the machine for one of their regular products. At any rate, much of the US plant that was surplus to requirements in 1945 was snapped up by contractors eager to make use of state-of-the-art equipment. One crawler dozer produced in the UK, which to an extent surpassed the Caterpillar D8, was the Vickers VR 180 Vigor that appeared in 1955. It was powered by the same Rolls-Royce powertrain that was fitted in the company's tanks, and it proved faster than the Cat D8. The Vigor had

The Ida Grove, Iowa-based Gomaco Corporation makes a range of pavers and trimmers for the road, airport runway and canal construction industry. This is a 9500 trimmer at work on a new road-building project, featuring extended sensor arms that help keep the machine on track.

ROAD ROLLERS, COMPACTORS & TARMAC LAYERS

a different track layout to the Caterpillar, with independent bogies that produced a smoother ride. It was also possible to change gear on the move. By the mid-1960s, however, the Vigor was out of production and Caterpillar, Allis Chalmers and Euclid were already more prominent.

TOOLS OF THE TRADE

In the chapter on motor scrapers, we see how they are productive machines whose primary role is to move material in a fast and efficient manner. Unlike a dump truck, they are able to put material into the profile needed, unaided by other machines to a very large extent. A dump truck can only deposit material in a heap and the operator or contractor is dependent on other machines to deal with the material. A scraper pretty much places the material where it is going to stay. On road contracts, the first job a scraper has to do is to remove the topsoil from the line of construction and to place it in heaps that can later be accessed, so that the topsoil comes back onto the job to cover the new embankments, cuttings and landscaped areas.

After that has been done, the serious work of bulk earth-moving commences, which is the creation of cuttings and embankments. The material generated from making cuttings is transported to the areas that will be embankments, forming a mean level between the two. Designers of roads try to balance the volume of material dug from cuttings with the requirement of material in the embankments, so that there's not too much material left over or, worse, not enough to play with. A reasonable surplus is the desirable result, which can be put into service to make up deficits in material that may be deemed unsuitable for construction use. Any final surplus can be lost in landscape areas, the form of which can be modified to accommodate any extra material.

The profile of the road allows no latitude in this area, as its design is fixed down to the last millimetre, as is all the supporting ancillary network such as the drainage system, the road gantries, the bridges, slip roads and underpasses. The new road has to interact or merge harmoniously with everything in its path, that's to say rivers or canals, other road systems, railways, electricity pylons, underground utility services including pipelines and cables and many other underground and topographical features. So, once decided on

The whole gamut of earth moving equipment is at work on a new road project, as scrapers and dozers first remove topsoil, and bulk earth moving follows during the creation of cuttings and embankments. After that, the compactors move in, and pavers and finishers complete the job.

ROAD ROLLERS, COMPACTORS & TARMAC LAYERS

A Bell 2406C 4x4 tractor unit with Mercedes-Benz power pulls an impact roller along a road construction project in Botswana. As with the sheep's-foot roller, the nodules on the wheels consolidate the material, while the roller frame carries extra weights to assist with surface compression.

and approved, the design of the road is fixed. Landscape features that are outside the road-line can remain flexible right up to the opening of the road.

After the blades and the scrapers have had their way with the landscape, the wheeled compactor may put in an appearance. This device looks like a cross between a road roller with rubber tyres at the back, a dozer blade on the front, and most notably, a vibrating steel drum in front. This can be smooth for dealing with sand and other fine-grain substances, or it can have a 'padfoot' drum instead, which has striations akin to a continuous sprocket, useful for mashing soils with a heavy clay content. A variation is the full-blown padfoot compactor, which has one of these padfoot drums, resembling an implement of medieval torture, at all four corners as well as a dozer blade at the front. As you'd expect, this is for really serious compacting, and has little to do with the finesse of the machines that ultimately create the finish of the road surface.

In the process of making a road, any mineral or organic material that's been dug up by the blade or scraper and dumped somewhere else will contain air voids. In muck-shifting terms, material placed in embankments, undulations and abutments must have the voids taken out so that it has the maximum structural strength. Some materials are suitable for construction, but others are not, and that's a matter for the geology department. Of those materials that are suitable, the best are the ones that have a variety of different-sized particles in their make-up. These compact easily, and bind well. Sand is suitable for construction, but as all the particles are the same size, it is incredibly difficult to compact. Sand is often compacted hydraulically, which means it gets sprayed with copious quantities of water as well as being rolled.

A 'Type-1' sub-base is easy to compact, as it contains chips, lumps and sands of various dimensions. Crushed material, once compacted, will rarely move so long as it contains sufficient 'fines', or sand. River-gravel will never compact as the particles are round, and therefore they won't bind.

The original compactor was the dead-weight steam or diesel roller, which worked by simply exerting pressure on the material that it was travelling over. This was superseded by the vibrating roller, which is a far more efficient machine. The vibrating roller had eccentric weights fitted inside the roller drum which was driven round at high speed by an engine. These compactors were either towed units or independent ride-on machines. Both types ranged in size from a pedestrian job to quite large machines. The original steamrollers were like galleons ambling majestically along. You just knew – like Kevin Kline in *A Fish Called Wanda* – that nothing would stand in their way. Kline's enigmatic 'survival' was perhaps due to the fact that the concrete wasn't set.

SHEEP'S FOOT ROLLERS

In earth-moving applications there is another variety of dead-weight roller, and this is the 'sheep's foot roller'.

ROAD ROLLERS, COMPACTORS & TARMAC LAYERS

A Witgen 2000 DC cold milling machine skims the top off a redundant road surface near Kyoto, Japan, prior to resurfacing. The mangled asphalt is dispatched up a conveyor into a tippper truck that moves slowly ahead of the miller.

Instead of the drum being smooth, it is fitted all over with blunt spikes that punch the material into itself. Again, both the towed variety and the self-propelled type are available. With vibrating rollers, the self-propelled version is normally called a compactor. The Caterpillar 815 and 825 models are good examples of a sheep's foot compactor, and the following description applies to both of them as they are identical apart from their dimensions. As we've come to expect with numerical designations, the higher the figure, the bigger the capacity, and the Cat 825 is the larger of the two. Typically, it deals with fill-areas that are being served by scrapers. The 815 is essentially a loading-shovel with the shovel equipment removed. Instead, it is fitted with a light bulldozer blade at the front. Rather than having large rubber tyres, it is fitted with four sheep's foot rollers as wheels, which are of a smaller diameter than the tyres on the equivalent loading shovel. It works by running backwards and forwards at as high a speed as it can, hammering the material underneath it. The little blade is present simply to skim the surface and level it out where necessary, but not for any serious pushing. The most work it does is to push material into those awkward corners that the big scrapers can't get into.

In earth-moving applications, the self-propelled vibrating rollers that are normally used have the roller at the front and the engine at the back, with rubber wheels either side of the engine. Normally on these machines, both the rubbers and the drum are driven. All-roller machines, that's to say, twin-roller and the traditional steam or diesel roller which has three rollers, are normally used to compact newly laid black-top. They're fitted with spray-bars over the rollers to keep them wet, so that they don't pick up the tar.

A further type of roller that's fitted with rubber-tyre shod wheels is produced by a French-based company called Alberet. Compactors of this type are quite heavy, and are fitted with five wheels across the front of the machine and six at the back. The principle is the same as that of the wobbly-wheel roller mentioned earlier, in that the back wheels compact the material that the front wheels miss because of the gaps between the wheels. The tyres are smooth, and the pressure of the tyres can be altered from the cab for the optimum compression for the material being compacted.

PAVING THE WAY

Once the groundwork has been done by the heavy plant, it's time for the road surface to be built. In the 1990s this was still normally composed of either asphalt or concrete, and indeed if a canal was being constructed, that too would require an impervious lining made of concrete.

An important piece of US kit that was imported into Europe in the war years for laying airfield runways and roads was the finisher or paver. Although a wide range of machines evolved subsequently to carry out paving applications on roads, canals, or airport runways, the concept changed little, and a 1940s paver is similar in configuration to one from the 1990s. Size was the main difference, and they varied in stature from hand-operated models to giant concrete trimmers that could lay 45.7-m (150-ft) wide strips. The paver performs two basic functions: one is to download the medium onto a prepared roadbed, and the

ROAD ROLLERS, COMPACTORS & TARMAC LAYERS

second is to level it. Manufacturers of pavers have included Blaw-Knox, Gomaco and Barber-Greene. The largest pavers available in the 1990s were the CMI SF 7004 and the Gomaco GP-4000. Both were capable of laying concrete to a depth of 45cm (18in) in a strip 15.24m (50ft) wide. They tipped the scales at over 44 tons, were 3.66m (12ft) high, and powered by 450bhp Cummins diesel engines. A more complex device for forming kerbs and gutters was Gomaco's CG5 Commander III, which was equipped with a side-shifting trimmer that enabled it to make tight-radius concave corners.

The basic paver consisted of a tractor unit integral with a chassis-mounted hopper, into which hot asphalt was tipped from a truck. The working temperature of the material had to be between 121 and 190 degrees centigrade. A couple of augers then fed the asphalt on a conveyor to a jigger, on which it was levelled and tamped out before being passed on to the screed unit. This was a large flat plate that was as wide as the section of road being surfaced, and widths of modern versions vary between 1.2m (4ft) and 12.2m (40ft). An operator perched on the top of the paver dispensed a flat, hot swathe of asphalt via the screed onto the roadbed in a process that continued until the asphalt hopper was empty. The depth of the carpet of tar was between 10.2cm (4in) and 30.4cm (12in), and the machine progressed at a speed that was measured in feet per minute – somewhere between 60ft and 150ft per minute. A number of workmen armed with rakes, known as screed men, accompanied the paver to ensure the depth was consistent. More modern machines incorporated a vibrating mechanism that extinguished any air bubbles, while depth and heat of the mix were regulated by automatic controls. In the post-war era, machines like the Barber-Greene finisher transformed roadbuilding and resurfacing worldwide. Consistency was the keynote with a paver, and in the 1990s, one would set a contractor back $250,000.

FLAT AS A PANCAKE

In the early days, the ubiquitous steamroller wasn't far behind the paver. More specialised compactors followed in its wake, generally of three different types. First of the finish compactors was known as a breakdown roller, which

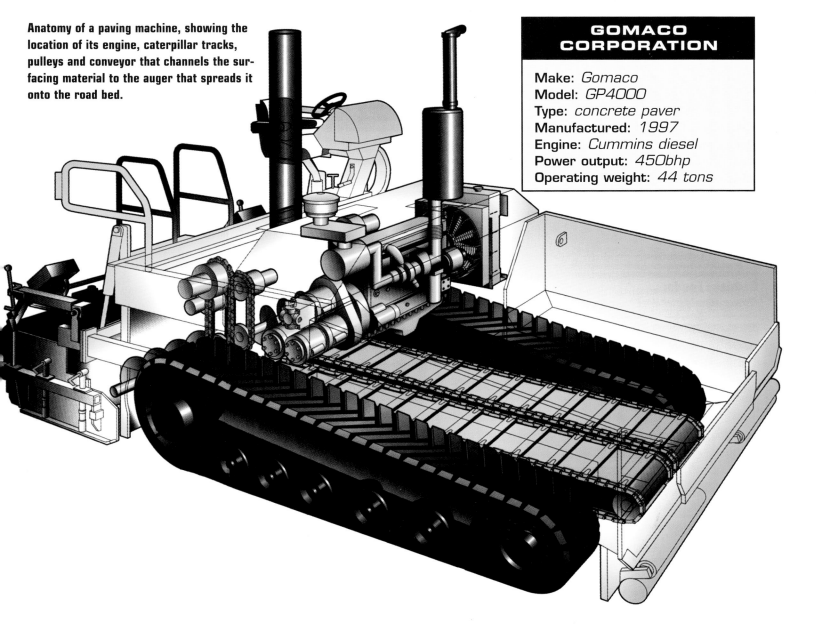

Anatomy of a paving machine, showing the location of its engine, caterpillar tracks, pulleys and conveyor that channels the surfacing material to the auger that spreads it onto the road bed.

GOMACO CORPORATION

Make: *Gomaco*
Model: *GP4000*
Type: *concrete paver*
Manufactured: *1997*
Engine: *Cummins diesel*
Power output: *450bhp*
Operating weight: *44 tons*

mashed the asphalt and compacted it with two vibrating steel drums. The first pass of this machine was known in the road-building fraternity as pinching the joint. It was followed by a second roller fitted with nine large pneumatic tyres, which had the effect of further compressing the asphalt, and its tyres helped seal the surface. The third machine to arrive on the scene was the finish roller, equipped with a pair of smooth, non-vibrating steel rollers, which made good the final sealing of the road surface.

One of the longest established manufacturers of compacting equipment was HAMM, founded in 1878 by Franz and Anton Hamm under the company name Maschinenfabrik Gebruder HAMM, based at Tirschenreuth, Germany. In 1911 Hans Hamm designed and built the company's first diesel-driven road roller. In 1999, HAMM was the oldest manufacturer of compactors still active in Germany.

By the late 1920s, the increasing demands for compaction of the metalled road network led HAMM to concentrate exclusively on the development and manufacture of road rollers. In 1932, HAMM patented the world's first tandem roller with all-wheel drive and all-wheel steering and created a minor revolution in the compactor construction industry. Surviving the turmoil of the Second World War, HAMM's compactors in the 1950s were aesthetically attractive and rounded in aspect. Export sales soared from 1953 onwards.

A new plant came on stream in 1955 at Bahnhofstrasse, Tirschenreuth, and HAMM became a serious mass producer of compactors, with the HAMM Eagle as its trademark. All this time, the patent meant that from 1932 up to 1970 HAMM was the only manufacturer in the world entitled to build tandem rollers with all-wheel drive and all-wheel steering, with the exception of a licence granted to Japan in 1964. With the boom in long-haul road construction, this type of compactor was used by roadbuilders all over the world.

In 1963, HAMM released a rubber-tyre shod wheeled roller with all-wheel steering and all-wheel drive. It was also the first construction machine in the world with hydrostatic drive, and during the construction of the highly significant Rhine-Main-Danube canal that linked two great European waterways, several models of this type were used extensively. They were mainly the tyre-shod GRW 10 and GRW 15 rollers, which remained virtually unchanged into the late 1990s.

By 1980, the road-building industry had developed special surfaces and coatings that required increasingly sensitive compacting. HAMM responded by introducing a new oscillation method of compacting. Unlike normal vibratory rolling, which compacts the material by vertical vibrations, the oscillating method keeps the compacting drum in permanent contact with the ground. The transmission of horizontal rather than vertical forces into the ground yields much better compaction results and, at the same time, is less stressful on the operator, material and the environment. The new HAMM oscillatory rolling machine was unveiled in 1989, after years of field trials. This was followed in 1992 by a fully hydrostatic soil stabiliser and recycler, which was the largest of its type in the world. By 1999, HAMM had come up with a new articulated tandem vibration and combination roller line to augment its established DV line.

THE PRANCING HORSE

The road-roller manufacturer with the longest record in the UK is Aveling Barford. Based at Grantham, Lincolnshire, and owned by Wordsworth Holdings Plc, Aveling Barford's range of hydrostatic-drive road rollers in 1999 included the HDC 12, HDC 13, HDC 14, and HDC 15. Two were fitted with cabs, and two were not, and the basic cab-less models, designated DC 12/13 and DC 14/15, were not available with lights. All four were un-ballasted ex-factory, and water was the ballast medium. Biggest of the four machines, the HDC 15, took its power transfer from its Perkins engine through an Eaton hydrostatic pump to twin Poclain hydraulic motors that served the rear rolls. This efficient and reliable transmission system gave smooth, infinitely variable speed changes through 0 to 9.8km/h (0 to

Prior to being resurfaced, the blacktop is removed from an old road by a milling machine. This is the undercarriage of Wirtgen's W-2000 cold milling machine, which has hydraulically-operated swivelling crawler tracks and special automatic transmission that co-ordinates the milling and driving gear.

Another variation on the deadweight roller is this Aveling-Barford VibraVista finish compactor, which is shod with tyres on its driven axle and a vibating steel drum at the front. The compression is transmitted through the frame surrounding the drum.

6miles/h). The smoothness of this system was particularly effective for cutting and jointing operations, and an electronically actuated differential lock that was exclusive to Aveling Barford's HDC Series machines optimised the drive on slippery surfaces, to give the driver more control and to produce a better finish. An 80-litre hydraulic tank was fitted for the cooling system. Access to the cab was via a non-slip ladder with hand rails and interior grab handles, while rubber-cushioned mountings to the cab frame minimised vibration and reduced noise levels. Roll-over protection was certified to 15 tons. In the cab, a universally adjustable seat was fitted to recessed tracks, which allowed the operator to traverse the full width of the cab, and afforded him excellent visibility down to the rolls from a seated position. Forward and backward, high or low position adjustment was also available. Power steering provided five turns left and seven turns right, via a steering wheel that was column adjustable for height and rake. Internal environment controls included fresh air/heater/demister facility, and front and rear windshield wash/wipe, all located in an overhead console together with a radio/cassette player fitted as standard equipment. The roller's controls and dashboard included twin pods at both sides of the cab, and dual emergency switches activated the fail-safe braking system that brought the rollers to a dead stop. All controls were hand-operated, with no floor controls at all, leaving the cab clear for stowage. The HDC 15 was powered by an 82bhp 4-litre Perkins 1000 series diesel engine, rated as one of the quiet-est and cleanest engines in its class, complying with mandatory EC regulations for noise and exhaust emissions. Service points were accessed via lockable servicing doors. A range of optional equipment included a hydraulic three-tine scarifier, independent hydraulic asphalt edge cutter, hand pump and hose, spring scrapers and slinging plates for roller cleaning, and cocoa mats for better traction.

SET IN CONCRETE

The concrete paver performed much the same function as the asphalt paver described above. The main difference between the two was that temperature control wasn't critical with concrete. It just needed to be moist. It was unloaded into the paver's hopper by dump truck or concrete carrier, and then spread by means of an auger across the ribbon of roadbed. A more recent innovation was the dowel bar inserter, which offloaded steel reinforcing rods as the machine progressed, previously a labour-intensive task that was performed by hand by workmen. The spreader was followed up by the screed or finisher, which levelled the concrete and extracted surplus material. Some pavers had more than one screed, and some were fitted with a vibrating mechanism that consolidated the concrete still further. When curves were required, slipform pavers were used. These did away with the hand-formed shuttering originally used to make up bends in the road, and were capable of forming kerbs and gutters of varying profiles and configurations.

The heyday of the giant paver was in the 1950s and 1960s during the construction of the US interstate highway system. The original surfaces had a finite life, and some thirty to forty years on they needed to be resurfaced, and the giant pavers were back in business. Other major projects where these machines came into their own was the construction and rebuilding of airport runways, bridges, and laying the bed and sides of canals and waterways. These finishers were capable of laying swathes of concrete 45.7m (150ft) wide. They were in effect low bridges with a lattice work gantry that stretched across the canal bed and a set of bogies on either side. They follow on from concrete carriers or dump trucks that deposit their loads onto the designated strip, and a long cylinder beneath the finisher levels and compacts the surface. These machines are sufficiently versatile to be able to form the sides of the channel as well. One such task was the construction of the Coachella Canal in California in 1988. The paver was specially built by Gomaco and measured 31.4m (103ft) wide across the top, tapering to 14.6m (48ft) at the bottom, which corresponded more or less with the shape of a ship's hull.

These machines demonstrated that techniques of road construction had certainly come a long way since the days of Telford and McAdam, and certainly since the Romans.

INDEX

ACCO
 Dozer 75
 graders 101
J.D. Adams, graders, Little Wonder 17-18, 99
airport construction
 Hong Kong 8
 WW II 166-7
Åkerman 12
 excavators 48, 49
Alberet, compactors 170
Allis-Chalmers
 see also Fiat-Allis
 bulldozers, HD-series 70, 71
 scrapers 90
Armstrong, Sir W.G. 35
Arrol, dock-side crane 147
Australia
 see also Tiger
 coal mining 31, 43, 128-9
Aveling & Porter
 see also Aveling-Barford
 compactors 173
 company history 163-4
 diesel rollers 162, 172-3
 steam rollers 163-4
Aveling-Barford
 dump trucks 134-5
 graders 96, 99, 101
 road rollers 163, 164, 167

Bamford, J.C. 15
 see also JCB
Barford & Perkins, road roller 165
Bauer, drilling rigs 63
Beck & Pollitzer, tower cranes 148
Bell
 dump trucks 134
 graders 98
 loaders 14
 tractor/impact roller 169
Bladerson, bulldozers 16
Boulet, bucket-wheel excavators 20
Buckau-Wolf, bucket-wheel excavators 38
Bucyrus
 see also Bucyrus-Erie; Ruston-Bucyrus
 company history 12, 27, 28-9
 Marion taken-over 42
 stripping shovels 10, 19
Bucyrus-Erie
 drilling rigs 58, 59

excavators 29
loading shovel 30
walking draglines 10-11, 19, 32
 Big Muskie 33, 33
bulldozers 65-6, 73, 75
 amphibious 16-17
 dual working 72
 history 16, 69-79
 operation 66-9, 67
 wheeled 114
Byelorussia Automobil, dump trucks 134

Canada
 see also Doerr; Dominion Road; Sicard; Titan
 coal mining 16
 diamond mines 43
 oil sand quarries 42, 43, 45, 46, 48, 100
 open-cast mining 12
canal construction
 Coachella 22, 173
 Manchester Ship 10, 10-11, 27
 Panama 8, 11, 28
 Rhine-Main-Danube 172
 San Luis Dam 91
 Suez 8, 10, 38
Case, loaders 15, 108-109
Caterpillar
 bulldozers 16, 70, 71, 72, 73
 D9 series 64-5, 67-9, 70, 79
 D10 16-17, 70, 72-3
 compactors 170
 company history 13, 128
 crawler loaders 117
 dump trucks 118-19, 128, 135-6
  769 9, 128, 139-41, 139
  785 48, 135
  793-series 124, 136, 138
 excavators 34-5, 46, 46-8
  5230 46-7, 48
 graders 92-3, 96, 99-100, 99
  24H 18, 94, 101
  Auto Patrol 18, 99
 loaders 104, 106, 118-19, 140-41
  994 series 102-3, 106, 107, 108
 scrapers 87, 88-9, 90-91, 90, 91
  637 79-83
 wheeled dozers, 854G 116

Channel Tunnel 20, 22, 50-51, 56-7, 57-60, 71, 142-3, 149
Clark Michigan
 loaders 105
 wheeled dozers 116
CMI
 graders 100
 pavers 171
coal mining
 Canada 12
 shaft
  long-wall 52
  rock drilling 54
 U.S.A. 12, 28, 29, 33, 49
Coles
 cranes
  mobile 151-2
  telescopic 152-3
compactors 169-70, 172
Cowans-Sheldon, railway cranes 150
cranes
 barge-mounted 147, 159
 cable 154-5
 Channel Tunnel 149
 crawler 157-9
 dam construction 154-5
 derricks 147, 147
 dock-side 145-7, 145
 gantry 147
 goat-leg 19, 144
 history 143-7
 railway 149-50
 semi-submersible 61-2
 steam 144-5
 tower 22, 148-9
crawler transporters, Space Shuttle 6-7, 22

Daewoo, excavators 49
dam construction 119, 121, 154-5
Dart
 dump trucks 122-3, 130
 loaders 105-6
DDT, dump trucks 135
Demag 12
 see also Komatsu-Demag
 cranes 145, 155, 159
Dennis, fire engines 16
Doerr, bulldozers 75
Dominion Road, graders 100
Donson Equipment, dump trucks 16
draglines see walking draglines

dredging 31
Dresser see Komatsu-Dresser
 drilling rigs 63
 see also oil rigs
drivers
 health 23
 road rollers 165
 training 23
dump trucks
 articulated 128
 bottom-dump 128-9, 130
 colours 122
 electric/diesel-electric 125-7
 gas turbine 129-30
 history 15-16, 119, 121-3
 operation 139-41
 tyres 134
Dunbar, steam crane navvy 27
Dynapac, compacting roller 165

Erie Steam Shovel Co 29
Euclid
 see also Euclid-Hitachi
 bulldozers 70-71
 dump trucks 121-2, 138
  gas turbine 129-30
  TracTruk 16, 121
 scrapers 90
Euclid-Hitachi
 dump trucks 122
 excavators 48
excavators
 bucket wheel 20, 37-8
 crawler 38
 drag line 30-31
  walking 11-12, 19-20, 31-2
 gold dredges 39
 history 9-11, 26-8
 hydraulic operation 39-40
 vs cable 42, 43
 operating 40, 42
 trenchers 38

Fiat-Allis
 bulldozers 71
 scrapers 78
Foden, dump trucks 121
France
 see also Alberet; Boulet; Liebherr; Proclain; Quadral; SICAM
 Channel Tunnel 20, 22, 50-51, 56-7, 57-60, 71
 Le Havre docks 146

Mont Cenis tunnel 53

Gallion, graders 14
Gannon Manuf. Co., scrapers 18
General Motors
   see also Euclid; Terex; Titan
Germany
   see also Atlas; Bauer; Buckau-Wolf; Demag; Klemm; Komatsu [Germany]; Krupp; Lubecker; MAN; O&K; Tadano Faun
gold
   dredges 39
   mining 21
Gomaco
   pavers 19, 22, *160-61*
      GP4000 171, *171*
   trimmers 22, *166-7*
graders
   controls 94-6, *100*
   history 14, 17-18, 99-100
   operating 95-9
Grove [Cranes *later* Worldwide]
   cranes 153-4
      off-road 155-7, *156*
   platforms 153-4

Halla
   excavators 49
   loaders 113
HAMM, compactors 172
harbour construction, cranes 144-5
Heathfield, dump trucks 135
Hitachi *see* Euclid-Hitachi
Holt, Benjamin 13, 18
Holt Bros
   bulldozers 69
   steam tractors 128
Hough
   see also International Hough
   loaders 15, 104
Hough, Frank G. 15, 104
hydraulic operation
   development 35
   excavators 39-40

Ingersoll Rand, drilling rigs 63
International Harvester
   bulldozers 66, 70
   dump trucks 124
   scrapers 90
International Harvester [Poland], bulldozers 75
International Hough
   loaders 104, *104*, *105*, 106
   wheeled dozers 116
Italy *see* ACCO; Fiat-Allis;

Massarenti

Japan
   see also Hitachi; Kato; Kobelco; Komatsu; Mitsubishi; SMEC
   road construction *170*
   Seikan Tunnel 22
JCB
   company history 36
   Dancing Diggers 9
   excavators 8-9, 13, 34, 36
      JS-series 36, *36*, 39
   loaders *112-113*
John Deere, bulldozers 16
Joy Mining
   company history 53
   conveyor trains 53
   long-wall shears 53
   shuttle cars 52, *52*

Kato
   cranes 155
      off-road 157
Kawasaki, loaders *110*
Klemm, drilling rigs 63
Kobelco, excavators 49
Komatsu [Demag *later* Germany], excavators 43, *43*, 45-6, 46
Komatsu [*later* -Dresser]
   bulldozers 16, 67, 72, 73-4, *74*, 75
      D375A series *68-69*, 74-5, *75*
   dump trucks *15*, 131, 136
   graders *95*, 97
   loaders 107, 108, *111*, *114*
Korea *see* Daewoo; Halla; Samsung
Kroll, tower cranes 22, *148*, 149
Krupp
   bucket-wheel
      excavators 20, *21*, *24-5*, *36-7*
   graders 27
   deck cranes *157*
   drilling rigs 63

Leonardo da Vinci 38
LeTourneau
   see also LeTourneau-Westinghouse; Marathon LeTourneau
   bulldozers 70
   dump trucks 125, 133
      electric/diesel-electric 125
   graders 100
   loaders 109
      L series 109-10, *111*-12

   scrapers 14, 86, 87-8
      LT-360 17, *82-3*, 87-8
      Mountain Mover 18, 86
   wheeled dozers 114-15
LeTourneau, Robert Gilmour 14, 69-70, 86, 109, 125
LeTourneau-Westinghouse
   dump trucks 124
   scrapers 89
Liebherr
   bulldozers *16*, 75
   cranes
      crawler *158*
      mobile *152-3*, 155
   drilling rigs 63
   dump trucks
      T252 *123*, 135
      T282 136-8, *136-7*, 138-9
   excavators *40-41*, 42, 46, *123*
loaders
   wheeled
      history 14-15, 104-5
      operation *111*, 115
      tyres 105
Lubecker, bucket-wheel excavators 20

Mack
   dump trucks 123-4
   AC Bulldog 15, 119, 121
MAN, fire engines 16
Marathon LeTourneau, loaders 110, 111
Marion
   company history 12, 27-8, 130
   dump trucks 130
   excavators 42
   stripping shovels 19, 28, 32
   walking draglines 20
   wheeled dozers 116
Massarenti, drilling rigs 63
McAdam, John Loudon 162-3
E.G.Melroe, loaders, Bobcat 15
Menzie Muck, backhoe 13
mining shovels
   cable 12
   hydraulic 12-13
Mixermobile, loaders 104-5
Monighan Machine Co, walking draglines 19-20, 31
Moxy, dump trucks *127*, 128

navvies
   human 11, 26
   steam 27
New Holland, graders *97*
NKMZ, walking draglines 20

Northfield, dump trucks 128
Northwest 12

O&K
   dump trucks *126*, 135
   excavators 13, 44-5
      bucket-wheel 20, 38
      RH400 *13*, *44-5*, 45
   graders 100
   loaders 113
oil rigs
   land 22
      history 60-61
   North Sea 22, 62, *62*
   South Africa 61-2
oil sand quarries, Canada 42, 43, 45, 46, 48
operating
   bulldozers 66-9, *67*
   excavators 40, 42
   scrapers 79-85, 86, *86*
Orenstein & Koppel *see* O&K
Otis, steam shovel 26

P&H
   see also Joy Mining
   excavators 29-30, 42-3, *49*
   walking draglines 31
      Ace of Spades 32, *32*
Page, walking draglines 20
pavers 19, 166-7, 170-71, 173
Pawling & Harnischfeger *see* P&H
Peerless
   dump trucks 130
   wheeled dozers 116
Peterson, scrapers 91
Poclain
   drilling rigs 63
   excavators 36
Poland *see* International Harvester [Poland]
power
   computer controlled 36
   diesel 12, *85*, 125
   diesel-electric 106, 109, 125-7, 130-31
   electricity 28, 29-30, 125
   gas turbine *122*, 129-30
Priestman & Smith, excavators 12, 30

Quadral, cranes 156

railways
   construction 9, 26-7, 28
   cranes 149-50
Ransomes & Rapier 12

walking draglines 20, 32
RayGo, graders 100
RB Variable Hydraulic, excavators 36-7
Rimpull, dump trucks *129*
road construction 18-19, 78-9, 139-41, 165-7
  see also compactors; graders; pavers; road rollers; scrapers
  equipment 168-9, *168*
  history 14, 17-18, 77, 161-6
Robbins, tunnel borers *50-51, 57-8, 61, 63*
Russell
  bulldozers 72
  graders 14, 18, 99
Russia (*formerly* Soviet Union)
  see also Byelorussia Automobil; NKMZ; UZTM
  Bucyrus licence 28-9
  tunnel borers 53-4
  coal mining 130
Ruston & Dunbar
  excavators, steam 10
  steam shovel 27
Ruston & Hornsby, excavators 30, 31
Ruston & Proctor, excavators 10
Ruston-Bucyrus
  excavators 30
  walking draglines 20

Samsung, excavators 49

T.G.Schmeiser, scraper 18
scrapers
  history 14, 18-19, 86-7
  laser guidance 19
  operation 79-83, *80*, 86, *86*
ships' cranes [derricks] 147, *147*
Sicard, loaders 106
SMEC, loaders 107-8
South Africa
  see also Bell
  copper mines 125
  Mossgas offshore exploration 61-2
Space Shuttle, crawler transporters *6-7*, 22
Stothard & Pitt
  company history 144-5
  cranes
    block setting 144
    dock-side 145-7, *157*
stripping shovels 19, 28
Stubbs, Gerald
  bulldozer operation 66-9
  dump truck operation 139-44
  excavator operation 40, 42
  scraper operation 79-85
Sweden *see* Åkerman; Volvo
Switzerland *see* Menzie Muck

Tadano Faun, off-road cranes 156
Terex
  dump trucks 16, *120-21*, 132, *132-3*, 133, *135*, 138

scrapers *15, 76-7, 79, 81, 84, 91*
Tiger, wheeled dozers 116-17, *116*
Titan, dump trucks 33-19 133
transmission systems
  chain 123
  diesel-electric 106, 109, 125-7, 130-31
tunnel boring 21-2, 51-2
  Channel *20*, 22, *50-51, 56-7*, 57-60
  cutter head 54, *55*
  full-face 54-6, *56-7*, 57, 58-9
  history 53-4, 57
  manual 53
  Mont Cenis 53
  Seikan 22
tyres
  dump trucks 134
  loaders 105

Unit Rig
  dump trucks *122*, 125-7, 129, 130, *131*
  gas turbine *122*, 129
U.S.A.
  canal construction 22, 91
  coal mining
    open-cast 19, 28, 29, 33, 49, 128-9
    shaft 52
  copper mining 122
  dam construction 119, 121
  oil exploration 60-61

railroad construction 28
road construction 18, 87, 99
UZTM, walking draglines 20

Vermeer, trenchers *12*, 38-9
Vickers, crawler dozer 167-8
Volvo
  see also Åkerman; Euclid-Hitachi
  dump trucks 128
  excavators 49
  loaders 113-14

WABCO [Westinghouse]
  dump trucks 130-31
  loaders 105
  scrapers 89-90
  walking draglines, history 11-12, 19-20, 31-2
Western
  dump trucks 16, 124
  wheeled dozers 115
wheeled
  compactors 19
  dozers 114
Whitaker Steam Shovel 9, 27
J.H.Wilson & Co, excavators 10-11
Wirtgen, cold miller 170, 172
Wiseda, dump trucks 135
World Wars, Second 18, 26, 70, 73, 78, 87, 166-7

PICTURE CREDITS

Associated Press: 8
Bell Articulated Dump Trucks: 14
Bell Equipment: 98, 134, 169
Brookes & Vernons: 112-113
Bucyrus: 10, 30, 33, 58, 59
Builders Group Library: 9 (Anthony Honess), 15 (t) (Anthony Honess), 64-65 (Anthony Honess), 79 (Anthony Honess), 120-121 (Anthony Honess), 144, 150, 168
Case Corporation: 102-103, 108-109
Caterpillar: 48, 87, 90, 94, 106, 107, 116, 117, 118-119, 124, 138
Gomaco: 22, 160-161, 166-167
Grove Cranes: 156
International Harvester: 66, 104, 120 (t)
JCB: 36
Joy Mining Machinery: 52, 53
Kawasaki Heavy Industries Ltd: 110
Komatsu: 15 (b), 43, 67 (t), 68-69, 75, 95, 97 (b), 111 (b), 114
Krupp Materials Handling: 21, 24-25, 27, 36-37, 154-155, 155, 157 (t)
Liebherr: 16, 40-41, 42, 115 (t), 123, 136-137, 152-153, 158 (both)
Lincolnshire County Archives / Wordsworth Holdings Plc: 23, 96 (b), 99 (t), 101, 162, 163, 164, 173

Marshall Cavendish: 11, 16-17, 151
Moxy Trucks, Norway: 127
New Holland: 97 (t)
O&K: 13, 44-45, 126, 132
P&H Mining Equipment: 29, 31, 32, 49
QA Photos Ltd: 20, 46-47, 50-51, 54, 55, 56 (t), 60, 61, 63, 71, 74, 76-77, 81, 84, 92-93, 96 (t), 140-141, 142-143, 145 (b), 149
Rimpull: 129
Mike Schram: 72, 99 (b), 165
Shell Photo Services: 62
Terex Mining: 85, 91, 135
John Tipler: 19, 78 (b), 88-89, 125, 139, 145 (t), 159
TRH Pictures: 6 (NASA), 18, 26, 73 (US Marine Corps), 78 (IWM), 80, 82-83, 105, 122, 146 (both), 147, 157 (b)
Unit Rig, Tulsa: 131
Vermeer Manufacturing: 12
Wirtgen GmbH: 170, 172

ARTWORK CREDITS

Richard Burgess: 28, 34-35, 38, 39, 56-57 (b), 67 (m), 70, 86, 100, 111 (t), 115 (b), 128, 130, 148, 171
Wylde Parnelle / WBM: 45